MACMILLAN/McGRAW-HILL

Math
Test Prep
Practice
Workbook

Grade 5

The **McGraw·Hill** Companies

Macmillan McGraw-Hill

Published by Macmillan/McGraw-Hill, of McGraw-Hill Education, a division of The McGraw-Hill Companies, Inc., Two Penn Plaza, New York, New York 10121.

Printed in the United States of America

Contents

Lesson 1.1

PART 1 • Multiple Choice

Choose the best answer.

1. The girl below is 3 feet tall. Which is a reasonable estimate of the height of the man standing next to her?

   A. 4 feet   C. 6 feet

   B. 5 feet   D. 8 feet

2. Janet said, "My textbook is really heavy!" Which estimate is most likely the weight of her textbook?

   F. 5 pounds   H. 50 pounds

   G. 25 pounds   I. 250 pounds

3. The left pitcher contains 16 fluid ounces of lemonade. Which is the most reasonable estimate of the amount of lemonade in the pitcher to the right?

   A. 30 ounces   C. 50 ounces

   B. 40 ounces   D. 60 ounces

PART 2 • Short Response

Record your answers in the space provided.

4. Dan can stamp and seal 3 envelopes in one minute. What is the most reasonable estimate of the number of envelopes Dan can stamp and seal in 1 hour?

_____

5. The top pencil is 2 inches long. What is a reasonable estimate for the length of the bottom pencil?

_____

PART 3 • Extended Response

Record your answer in the space provided.

6. A store runs a contest to guess the number of pennies in a jug. You have the same jug at home with 500 pennies. Explain how you could use this information to estimate the number of pennies in the store's jar.

_____

_____

Lesson 1.2

PART 1 • Multiple Choice

Choose the best answer.

1. In which place is the 3 in the number 1,290,531,874?

   A. thousands

   B. hundred thousands

   C. ten millions

   D. ten thousands

2. What is the place value of the 7 in 7,295,503,002?

   F. 7 billion   H. 7 million

   G. 70 billion   I. 70 million

3. How is 30,088 written in expanded form?

   A. 3,000 + 80 + 8

   B. 3,000 + 800 + 8

   C. 30,000 + 80 + 8

   D. 30,000 + 800 + 8

4. How is 45 million, 80 thousand, 18 written in standard form?

   F. 450,800,018   H. 45,800,018

   G. 450,080,018   I. 45,080,018

5. How is 1,000,000,101 written in words?

   A. one billion, one hundred one

   B. one billion, one hundred and one

   C. one million, one hundred one

   D. one million, one hundred and one

PART 2 • Short Response

Record your answers in the space provided.

6. What digit is in the ten millions place of the number 5,102,967,438?

   _____

7. How is the number two hundred three million, fifty thousand, five hundred three written in expanded form?

   _____

PART 3 • Extended Response

 Record your answer in the space provided.

8. When people write checks to pay for purchases, they usually write the amount in numbers and in words. Suppose a person is buying a house and has to write a check for $128,750. Describe a procedure you would use to convert numbers to words.

   _____

   _____

   _____

Lesson 1.3

PART 1 • Multiple Choice

Choose the best answer.

1. The large square below represents the number 1. What number is represented by the shaded portion of the large square?

   A. 63.00   C. 0.63

   B. 6.3   D. 0.063

2. Each large square below represents the number 1. What number is represented by the shaded portion of the squares?

   F. 110   H. 1.01

   G. 11   I. 1.1

3. A large square containing 100 equal small squares represents the number 1. In this large square, 41 small squares are shaded. What number is represented by the shaded squares?

   A. $\frac{10}{41}$ C. $\frac{100}{41}$

   B. $\frac{41}{10}$ D. $\frac{41}{100}$

PART 2 • Short Response

Record your answers in the space provided.

4. What number is represented by the shaded portion of the large square?

   _____

5. In a 100-square grid, the number of small shaded squares represent the number 0.17. How many small squares are NOT shaded in this figure?

   _____

PART 3 • Extended Response

 Record your answer in the space provided.

6. Explain how a square grid can be used to model the number 0.28. Include a grid with your explanation.

   _____

   _____

Lesson 1.4

PART 1 • Multiple Choice

Choose the best answer.

1. Name the place of the underlined digit?

   28.93?

   A. hundredths

   B. tenths

   C. ones

   D. tens

2. Name the place of the underlined digit?

   135.09

   F. hundredths

   G. tenths

   H. ones

   I. tens

3. Name the place of the underlined digit?

   394.27

   A. hundredths

   B. tenths

   C. ones

   D. tens

4. Which decimal is equivalent to 2.4?

   F. 0.024   H. 0.24

   G. 0.240   I. 2.40

5. How is nine and twelve hundredths written in standard form?

   A. 9.012   C. 91.2

   B. 9.12   D. 912

PART 2 • Short Response

Record your answers in the space provided.

6. The sides of a square are fourteen and eight tenths centimeters long. How do you express this length as a decimal?

   _____

7. Rebecca needs 25.2 yards of border for her room. Write the word name and expanded form for the amount of border she needs.

   _____

PART 3 • Extended Response

Record your answer in the space provided.

8. Explain how to graph the number 7.8 on a number line. Include a number line with your answer.

   _____

   _____

   _____

Lesson 1.5

PART 1 • Multiple Choice

Choose the best answer.

1. Which number completes the number sentence correctly?

   23,870 >

   A. 23,871   C. 23,807

   B. 23,907   D. 24,800

2. Which number completes the number sentence correctly?

   1.06 < ■

   F. 1.006   H. 1.6

   G. 1.016   I. 1.060

3. Which set of numbers is ordered from greatest to least?

   A. 104,705; 104,750; 107,405

   B. 107,405; 104,705; 104,750

   C. 104,750; 104,705; 107,405

   D. 107,405; 104,750; 104,705

4. Which set of numbers is ordered from least to greatest?

   F. 0.003; 0.02; 0.021; 0.3

   G. 0.003; 0.02; 0.3; 0.021

   H. 0.02; 0.003; 0.021; 0.3

   I. 0.02; 0.003; 0.3; 0.021

5. What number DOES NOT lie between 478.402 and 478.42?

   A. 478.4   C. 478.409

   B. 478.41   D. 478.418

PART 2 • Short Response

Record your answers in the space provided.

6. Austin uses a digital thermometer that displays temperatures using two decimal places. What is the lowest temperature greater than 99° that this thermometer can display?

   _____

7. Tina's car has an odometer that displays miles driven using one decimal place. On a trip to visit her relatives, Tina drove less than 80 miles. What is the greatest number of miles her car odometer could have displayed?

   _____

PART 3 • Extended Response

Record your answer in the space provided.

8. Explain how you could find five numbers that lie between 6.803 and 6.804.

   _____

   _____

Lesson 1.6

PART 1 • Multiple Choice

Choose the best answer.

1. In their major league baseball careers, Eddie Plank had 327 wins, Christy Mathewson had 373 wins, Robin Roberts had 286 wins, and Whitey Ford had 236 wins. Who had the most wins?

   A. Plank
   B. Mathewson
   C. Roberts
   D. Ford

2. In a race, Alia had a time of 15.52 seconds, Beth had a time of 15.07 seconds, Krystal had a time of 15.4 seconds, and Rita had a time of 14.88 seconds. Who finished second?

   F. Alia
   G. Beth
   H. Krystal
   I. Rita

Use the table for problems 3–4.

Train Distances Traveled	
Name	**Distance**
Meteor	896 miles
Coast Glider	943 miles
Mountaineer	974 miles
Red Sunset	745 miles

3. Which train traveled the farthest?

   A. Meteor
   B. Coast Glider
   C. Mountaineer
   D. Red Sunset

4. Which train traveled the second greatest distance?

   F. Meteor
   G. Coast Glider
   H. Mountaineer
   I. Red Sunset

PART 2 • Short Response

Record your answers in the space provided.

5. Joanna is thinking of a money amount that contains two digits to the left of its decimal point and two digits to the right of its decimal point. Each digit is different and each digit is greater than 0. The sum of the digits is 16. What is the greatest amount Joanna could be thinking of?

   _____

6. A decimal number is between 0 and 1. Each of its four digits to the right of the decimal point is different and greater than 0. The sum of those digits is 12. What is the least number this could be?

   _____

PART 3 • Extended Response

Record your answer in the space provided.

7. Suppose you had to order four 20-digit numbers. Explain how the place-value system allows you to do this quickly.

   _____

   _____

©Macmillan/McGraw-Hill

Lesson 2.1

PART 1 • Multiple Choice

Choose the best answer.

1. $\begin{array}{r} 3{,}905 \\ +4{,}887 \\ \hline \end{array}$

    A. 7,782   C. 8,782

    B. 7,792   D. 8,792

2. $\begin{array}{r} \$10.00 \\ -3.79 \\ \hline \end{array}$

    F. $6.21   H. $7.21

    G. $6.31   I. $7.31

3. $2.7 + 2.07 + 2.77 =$

    A. 5.11   C. 7.54

    B. 6.44   D. 31.84

4. Renaldo wants to buy 2 pounds of chopped meat. A package at the supermarket is labeled 1.86 pounds. How much more than the package's weight is the weight that Renaldo wants to buy?

    F. 3.86 pounds

    G. 1.14 pounds

    H. 0.24 pounds

    I. 0.14 pounds

5. In four weeks of work, Tyler earned $85.75, $98, $76.50, and $75. How much did Tyler earn altogether?

    A. $163.98   C. $315.25

    B. $235.25   D. $335.25

PART 2 • Short Response

Record your answers in the space provided.

6. Dana bought a scarf for $12.71 including tax. She paid the cashier using two $10 bills. How much change should Dana receive?

    _____

7. $0.909 + 9.09 + 0.9$

    _____

PART 3 • Extended Response

 Record your answer in the space provided.

8. When the numbers 1.47 and 3.28 are added, first add the hundredths column: $7 + 8 = 15$. Then write down "5" instead of "15." Explain why it makes sense to write the "1" in a different column. Then find the sum.

    _____

    _____

Lesson 2.2

PART 1 • Multiple Choice

Choose the best answer.

1. What is 4,852.0186 rounded to the nearest hundred?

   A. 4,800   C. 4,852.02

   B. 4,852.019   D. 4,900

2. What is 15.894 rounded to the nearest tenth?

   F. 15.8   H. 15.89

   G. 15.9   I. 16.0

3. A children's hospital received grants of $195,000, $88,000, and $325,000. Which is the best estimate of the total amount of these three grants?

   A. $700,000   C. $500,000

   B. $600,000   D. $400,000

4. The Cedar Trail is 8.9 miles long. The Eucalyptus Trail is 3.06 miles long. Which is the best estimate of how much longer the Cedar Trail is than the Eucalyptus Trail?

   F. 12 miles   H. 6 miles

   G. 11 miles   I. 5 miles

5. Movie tickets cost $9.75 for adults and $5.50 for children. Which is the best estimate of the total cost for a family of 2 adults and 3 children?

   A. $25   C. $35

   B. $30   D. $50

PART 2 • Short Response

Record your answers in the space provided.

6. What is 7,498 rounded to the nearest thousand?

   _____

7. What is 0.3855 rounded to the nearest hundredth?

   _____

PART 3 • Extended Response

Record your answer in the space provided.

8. The height inside a highway tunnel is 12.25 feet. Andy is driving a truck whose height, rounded to the nearest foot, is 12 feet. Can Andy take this truck safely into the tunnel? Explain your reasoning.

   _____

   _____

   _____

   _____

   _____

Lesson 2.3

PART 1 • Multiple Choice

Choose the best answer.

1. The price of a bleacher seat was $8 in 1999. For each of the next four years, the price went up $1.50 each year. What was the price in 2003?

   A. $15.50   C. $12.50

   B. $14   D. $9.50

2. Nick fills a pool with water. The water level rises by 0.2 feet every 10 minutes. The water is at 2 feet at 10:00 A.M. What will the water level be at 11:00?

   F. 2.2 ft   H. 3.2 ft

   G. 2.6 ft   I. 5.0 ft

3. Carol opened a bank account with $500. Each month she deposits $50. How much does Carol have in the account at the end of 12 months?

   A. $1,100   C. $600

   B. $1,000   D. $562

4. At 6 A.M., the temperature was 64°. Over the course of the next 12 hours, the temperature fell 1.5° per hour. What was the temperature at noon?

   F. 58°F   H. 51.5°F

   G. 55°F   I. 48°F

5. A snail crawls 5 inches from home. From there it crawls 0.75 inches every minute. How far is the snail from home after 5 minutes?

   A. 5.75 inches   C. 8.75 inches

   B. 8.55 inches   D. 10.75 inches

PART 2 • Short Response

Record your answers in the space provided.

6. Richie's black locust tree was 2 feet tall when he transplanted it. It grew an average of 0.25 feet each month during its first 7 months in the ground. How tall was the tree, in feet, after 7 months in the ground?

   _____

7. The cost to enter a water park was $19.95 in 1998. Then the cost was increased $3 per year for five years in a row. What was the cost in 2003?

   _____

PART 3 • Extended Response

 Record your answer in the space provided.

8. Helene read 84 pages of a 600-page novel before the winter break. During the break, she read 40 pages each day. The break lasted 10 days. Explain two different ways you could find how many pages she has left to read at the end of the winter break.

   _____

   _____

Lesson 2.4

PART 1 • Multiple Choice

Choose the best answer.

1. What property is used to rewrite the sum?

   $47.1 + 0 = 47.1$

   A. Associative   C. Distributive

   B. Commutative   D. Identity

2. What property is used to rewrite the sum?

   $775 + (225 + 19) = (775 + 225) + 19$

   F. Associative   H. Distributive

   G. Commutative   I. Identity

3. What property is used to rewrite the sum?

   $3.9 + 9.3 = 9.3 + 3.9$

   A. Associative   C. Distributive

   B. Commutative   D. Identity

4. What property is used to rewrite the sum?

   $0 + (2.5 + 11.3) = 0 + (11.3 + 2.5)$

   F. Associative   H. Distributive

   G. Commutative   I. Identity

5. Which sum DOES NOT equal 100?

   A. $0 + 100$

   B. $(37 + 20) + 43$

   C. $(8 + 44) + (16 + 32)$

   D. $(9 + 29) + (41 + 31)$

PART 2 • Short Response

Record your answers in the space provided.

6. What number makes the number sentence below true?

    $+ 57.8 = 57.8$

   _____

7. What number makes the number sentence below true?

   $(3.9 + 0) + 5.6 = $ $+ (0 + 5.6)$

   _____

PART 3 • Extended Response

THINK
SOLVE
EXPLAIN

Record your answer in the space provided.

8. Explain how the Associative Property can help you find the sum $150 + (850 + 889)$ using mental math.

   _____

   _____

   _____

   _____

   _____

Lesson 2.5

PART 1 • Multiple Choice

Choose the best answer.

1. Add. Use compensation.

   74 + 58

   A. 122   C. 142

   B. 132   D. 144

2. Subtract. Use compensation.

   353 − 298

   F. 651   H. 55

   G. 57   I. 51

3. Add. Use compensation.

   23.7 + 9.6

   A. 33.3   C. 34.1

   B. 33.4   D. 34.3

4. Subtract. Use compensation.

   12.3 − 3.5

   F. 8.8   H. 9.55

   G. 8.00   I. 9.45

5. Add. Use compensation.

   32.8 + 6.4

   A. 42.2   C. 39.2

   B. 41.6   D. 38.6

PART 2 • Short Response

Record your answers in the space provided.

6. Lisa earned $34.85 and spent $9.90. How much did she have left?

   _____

7. Jack wants to solve 16.3 + 12.8 by compensation. If he adds 0.2 to 12.8, what does he have to do to 16.3 to find the correct sum? Explain.

   _____

PART 3 • Extended Response

 Record your answer in the space provided.

8. Explain how to use compensation two different ways to find the sum 578 + 293.

   _____

   _____

   _____

   _____

Lesson 2.6

PART 1 • Multiple Choice

Choose the best answer.

1. Add.

   $7,584 + 5,928$

   A. 13,402   C. 13,502

   B. 13,412   D. 13,512

2. Subtract.

   $5,700 - 3,238$

   F. 2,362   H. 2,472

   G. 2,462   I. 3,538

3. Add.

   $\$78.92 + \$54.19$

   A. $133.11   C. $132.11

   B. $133.01   D. $132.01

4. Subtract.

   $91.34 - 46.29$

   F. 55.15   H. 45.05

   G. 45.15   I. 44.05

5. Add.

   $698.3 + 721.5$

   A. 1,529.8   C. 1,429.8

   B. 1,519.8   D. 1,419.8

PART 2 • Short Response

Record your answers in the space provided.

6. What number makes the number sentence below true?

   $52.8 - \square = 28.3$

   _____

7. Find the sum. Tell which method you used.

   $8,294 + 3,960$

   _____

PART 3 • Extended Response

 Record your answer in the space provided.

8. Carlos had $100 and spent $49.92. Explain how to use mental math to find how much money he had left.

   _____

   _____

Lesson 3.1

PART 1 • Multiple Choice

Choose the best answer.

1. Multiply.

   6×700

   A. 4.2   C. 420

   B. 42   D. 4,200

2. Multiply.

   30×200

   F. 60   H. 6,000

   G. 600   I. 60,000

3. Multiply.

   $40 \times 2,000$

   A. 800,000   C. 8,000

   B. 80,000   D. 800

4. Multiply.

   $30 \times 20 \times 20$

   F. 12,000,000

   G. 1,200,000

   H. 120,000

   I. 12,000

5. Multiply.

   $30 \times 60 \times 100$

   A. 1,800

   B. 18,000

   C. 180,000

   D. 1,800,000

PART 2 • Short Response

Record your answers in the space provided.

6. What is the missing number?

   $5 \times 20 = 100$

   $50 \times 20 = \square$

   $500 \times 20 = 10,000$

   $5,000 \times 20 = 100,000$

   _____

7. How many zeros will be in the product of 5,000 times 400? Explain.

   _____

   _____

PART 3 • Extended Response

 Record your answer in the space provided.

8. Mr. Alban makes wooden rocking chairs and sells them for $200 each. If he sells an average of 50 chairs a year, how much money will he make over a 20-year period? Explain how to find the answer by counting zeros.

   _____

   _____

   _____

©Macmillan/McGraw-Hill

Lesson 3.2

PART 1 • Multiple Choice

Choose the best answer.

1. Which number sentence shows the Distributive Property?

   A. $12 \times (40 + 7) = (12 \times 40) + (12 \times 7)$

   B. $12 \times (40 + 7) = (12 \times 40 \times 7)$

   C. $12 \times (40 + 7) = (12 \times 40) \times (12 \times 7)$

   D. $12 \times (40 + 7) = (40 \times 7) \times 12$

2. 6×88

   F. 428   H. 508

   G. 488   I. 528

3. How can 7×904 be rewritten using the Distributive Property?

   A. $(7 \times 900) \times (7 \times 4)$

   B. $(7 \times 900) \times (7 \times 40)$

   C. $(7 \times 900) + (7 \times 4)$

   D. $(7 \times 900) + (7 \times 40)$

4. 3×865

   F. 2,595   H. 2,495

   G. 2,585   I. 2,485

5. 9×539

   A. 5,751   C. 4,751

   B. 4,851   D. 4,551

PART 2 • Short Response

Record your answers in the space provided.

6. 6×94

   _____

7. 9×827

   _____

PART 3 • Extended Response

 Record your answer in the space provided.

8. Show how the Distributive Property can help you find the product 8×209 using mental math.

   _____

   _____

   _____

Lesson 3.3

PART 1 • Multiple Choice

Choose the best answer.

1. Multiply.

   $$\begin{array}{r} 67 \\ \times\ 6 \\ \hline \end{array}$$

   A. 780   **C.** 402

   B. 642   **D.** 362

2. Multiply.

   $$\begin{array}{r} 518 \\ \times\ 24 \\ \hline \end{array}$$

   F. 12,432   **H.** 12,202

   G. 12,332   **I.** 3,108

3. Zoe loads 18 cartons onto an elevator. Each carton weighs 48 pounds. What is the total weight of the cartons she loads onto the elevator?

   A. 864 pounds   **C.** 432 pounds

   B. 764 pounds   **D.** 66 pounds

4. A website averages 8,250 hits each day. How many hits does this website average during a 30-day month?

   F. 257,500   **H.** 25,750

   G. 247,500   **I.** 24,750

5. $25 \times 678 \times 4$

   A. 6,780   **C.** 47,800

   B. 18,984   **D.** 67,800

PART 2 • Short Response

Record your answers in the space provided.

6. 314×9

   _____

7. Tim drives 258 miles roundtrip to visit his parents each weekend. There are 52 weekends in a year. How many miles does Tim drive in a year to visit his parents?

   _____

PART 3 • Extended Response

 Record your answer in the space provided.

8. The multiplication shown below has been started by a student. Since $7 \times 3 = 21$, shouldn't the 0 in the last row shown be a 1? Explain why or why not.

   $$\begin{array}{r} 8{,}483 \\ \times\ \ 74 \\ \hline 33{,}932 \\ 593{,}810 \end{array}$$

   _____

   _____

Lesson 3.4

PART 1 • Multiple Choice

Choose the best answer.

1. Which property of multiplication is demonstrated below?

   $5 \times 8 = 8 \times 5$

   A. Associative   C. Distributive

   B. Commutative   D. Identity

2. Which property of multiplication is demonstrated below?

   $323 \times (40 + 6) = (323 \times 40) + (323 \times 6)$

   F. Associative   H. Distributive

   G. Commutative   I. Identity

3. Which property of multiplication is demonstrated below?

   $23 \times (58 \times 11) = (23 \times 58) \times 11$

   A. Associative   C. Distributive

   B. Commutative   D. Identity

4. To multiply 87 by 99, Ana multiplied 87 by 100 and then subtracted 87 from her product. Which property did Ana use?

   F. Distributive Property of Multiplication over Addition

   G. Distributive Property of Multiplication over Subtraction

   H. Identity Property of Addition

   I. Identity Property of Subtraction

PART 2 • Short Response

Record your answers in the space provided.

5. $79 \times 74 = (79 \times 100) - (79 \times \blacksquare)$
   Which number completes the number sentence above correctly?

   _____

6. $5 \times (38 + 83) = (5 \times 38) + (5 \times 83)$
   What is the full name of the property demonstrated above?

   _____

PART 3 • Extended Response

 Record your answer in the space provided.

7. How are the properties of multiplication like the properties of addition learned earlier? How are they different?

   _____

   _____

   _____

Lesson 3.5

PART 1 • Multiple Choice

Choose the best answer.

1. Wim estimated the product
    567 × 825 by first rounding each
    factor to the nearest hundred. What
    was Wim's estimate of the product?

    A. 40,000   C. 400,000

    B. 48,000   D. 480,000

2. Estimate the sum 698 + 712 + 688
    by clustering.

    F. 1,800   H. 2,000

    G. 1,900   I. 2,100

3. Teresa bought groceries for $3.89,
    $0.68, and $0.89. She estimated the
    total cost by first rounding each price
    to the nearest dollar. What was
    Teresa's estimate?

    A. $3.00   C. $5.46

    B. $5.00   D. $6.00

4. Rob ordered 185 pairs of sneakers for
    his shoe store. Each pair costs $21.99.
    Rob estimated the total cost by first
    rounding each factor to its greatest
    place. What was Rob's estimate of the
    total cost?

    F. $3,800   H. $4,180

    G. $4,000   I. $4,400

5. Which sum uses clustering to
    estimate 23.9 + 24.7 + 26.3 + 28.3?

    A. 20 + 20 + 30 + 30

    B. 23 + 23 + 23 + 23

    C. 25 + 25 + 25 + 25

    D. 24 + 25 + 26 + 28

PART 2 • Short Response

Record your answers in the space
provided.

6. Estimate the product
    48.4 × 124 by first
    rounding each factor to
    the nearest ten.

    _____

7. Estimate the product
    4.5 × 27.35 by rounding
    each factor to the nearest whole
    number.

    _____

PART 3 • Extended Response

 Record your answer in the space
provided.

8. Scott usually estimates by first
    rounding factors to the nearest ten
    or the nearest whole number. Today,
    he needs to have 86 copies of a
    photograph made. Each copy will
    cost $0.19. Is Scott's method of
    estimating a good idea? Explain
    your thinking.

    _____

    _____

Lesson 3.6

PART 1 • Multiple Choice

Choose the best answer.

1. A customer orders 50 boxes of felt markers. Each box contains 14 markers. How many markers did the customer order?

   A. about 500 C. exactly 500

   B. about 700 D. exactly 700

2. The Quinn family receives a $600 tax credit for each of its four chidren. What is the total tax credit that the Quinn family receives?

   F. about $2,000

   G. about $2,400

   H. exactly $2,000

   I. exactly $2,400

3. A fish store sells bluefish by weight. The price is $2 per pound. A single bluefish ranges in price from $4 to $6. Allison needs 10 bluefish for her restaurant tonight. How much will this cost Allison?

   A. about $50 C. exactly $50

   B. about $20 D. exactly $20

4. Kim can read 40 pages in one hour. Kim started reading around 9:00 A.M. and finished reading around 11:00 A.M. How many pages did Kim read?

   F. about 440 H. exactly 440

   G. about 80 I. exactly 80

PART 2 • Short Response

Record your answers in the space provided.

5. A carton contains 144 ornaments. A store ordered 16 cartons. How many ornaments did the store order?

   _____

6. A school has 25 classrooms. Each classroom contains 22 desks. How many desks are there in all?

   _____

PART 3 • Extended Response

Record your answer in the space provided.

7. A newspaper article begins with the following paragraph:
   Before a crowd of 50,000, the Panthers defeated the Titans 31-28 tonight. With 14 seconds left on the game clock, Drew Addams kicked a field goal into a 30 mile per hour wind to seal the victory for the Panthers. Discuss whether you think exact numbers or estimated numbers are used in this paragraph.

   _____

   _____

Lesson 4.1

PART 1 • Multiple Choice

Choose the best answer.

1. Multiply.

   8.4
   × 8

   A. 68.2   C. 66.8

   B. 67.2   D. 64.2

2. Multiply.

   3.29×100

   F. 0.329   H. 32.9

   G. 3.29   I. 329

3. Multiply.

   3.74×6

   A. 22.44   C. 20.24

   B. 21.44   D. 18.24

4. Find the multiple of 10 that makes the statement true?

   $\square \times 0.038 = 38$

   F. 10   H. 1,000

   G. 100   I. 10,000

5. Select the statement that is true.

   A. $0.9 \times 3 < 0.2 \times 10$

   B. $0.8 \times 8 > 0.5 \times 10$

   C. $0.3 \times 7 > 0.4 \times 6$

   D. $0.9 \times 5 < 0.4 \times 10$

PART 2 • Short Response

Record your answers in the space provided.

6. A ticket to the movie theater costs $7.25. How much would 5 tickets cost?

   _____

7. A piece of candy costs $0.15. How much would 1,000 pieces of candy cost? Explain how to find the answer without multiplying.

   _____

   _____

PART 3 • Extended Response

Record your answer in the space provided.

8. What conclusions can you make about a product of any number multiplied by a number that is greater than 1? Include examples in your explanation.

   _____

   _____

Lesson 4.2

PART 1 • Multiple Choice

Choose the best answer.

1.   0.5
    $\times 0.5$

    A. 25   C. 0.25
    B. 2.5   D. 0.025

2. The large square below represents the number 1. Which product is represented by the shaded portion of the large square?

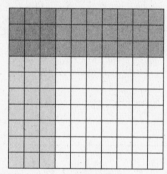

    F. $0.3 \times 0.3 = 0.09$
    G. $0.3 \times 0.3 = 0.39$
    H. $0.3 \times 0.3 = 0.9$
    I. $0.3 \times 0.3 = 9$

3. 0.7×0.8

    A. 0.56   C. 0.056
    B. 0.54   D. 0.054

4. A packet of rosemary weighs 0.2 ounce. Kevin used one-half, or 0.5, of the rosemary for cooking. How much of the packet did Kevin use?

    F. 0.1 ounce   H. 0.01 ounce
    G. 0.05 ounce   I.   0.005 ounce

PART 2 • Short Response

Record your answers in the space provided.

5. An Internet search engine claims it can find search results for you in 0.7 seconds. One of its competitors claims it can find results in one half, or 0.5, of that time. How many seconds does the competitor claim it will take for a search?

    _____

6. 0.9×0.9

    _____

PART 3 • Extended Response

Record your answer in the space provided.

7. The decimal number 0.5 can be read as "five tenths," a fraction that is equivalent to one half. Use this fact to help explain why the product of 0.5 and 0.8 is less than 0.8.

    _____

    _____

    _____

Lesson 4.3

PART 1 • Multiple Choice

Choose the best answer.

1. Multiply.

   $$\begin{array}{r} 3.47 \\ \times\ 2.4 \\ \hline \end{array}$$

   A. 8.328 C. 8.368

   B. 8.338 D. 8.428

2. Multiply.

   0.074×3.7

   F. 27.38 H. 0.2738

   G. 2.738 I. 0.02738

3. Multiply.

   9.42×4.6

   A. 43.342 C. 43.232

   B. 43.332 D. 42.332

4. Find the number that makes the problem true.

   $$\begin{array}{r} 4.6 \\ \times\ 3.1 \\ \hline 1\ \square.26 \end{array}$$

   F. 4 H. 2

   G. 3 I. 1

5. Which statement is true?

   A. $1.3 \times 1.2 < 0.6 \times 1.3$

   B. $2.1 \times 0.4 > 0.5 \times 3.4$

   C. $3.2 \times 1.4 > 2.5 \times 1.3$

   D. $0.8 \times 1.9 = 1.42$

PART 2 • Short Response

Record your answers in the space provided.

6. Bart earns $7.50 per hour. He worked 3.5 hours on Saturday. How much money did he earn on Saturday?

   _____

7. Find 0.5×0.3. Make a grid to show that your answer is correct.

   _____

PART 3 • Extended Response

 Record your answer in the space provided.

8. Sharon wants to estimate 234.86×8.9 before finding the actual answer. Explain which of the following would be the best estimate: 200×5; 200×10; or 300×10?

   _____

Lesson 4.4

PART 1 • Multiple Choice

Choose the best answer.

1. Multiply.

   398
   × 9

   A. 3,472   C. 3,572
   B. 3,482   D. 3,582

2. Multiply.

   6.8
   × 3.9

   F. 27.62   H. 26.52
   G. 26.62   I. 25.52

3. Multiply.

   819 × 7

   A. 5,093   C. 6,033
   B. 5,733   D. 6,093

4. Multiply.

   13.4 × 2.8

   F. 16.20   H. 37.52
   G. 29.12   I. 49.52

5. Multiply.

   0.54 × 2.59

   A. 1.3986   C. 1.3996
   B. 1.3988   D. 1.4006

PART 2 • Short Response

Record your answers in the space provided.

6. A pound of apples costs $1.38. How much would 5.5 pounds of apples cost?

   _____

7. Show how to use the Distributive Property to find the product of 895 × 7.

   _____

   _____

PART 3 • Extended Response

 Record your answer in the space provided.

8. Nancy makes $8.00 an hour at the pet shop. She worked 42.5 hours last week. Find how much she made last week using a calculator. Then, check your answer by using another method to solve the problem. Describe the method you used and explain why you chose it.

   _____

   _____

   _____

Lesson 4.5

PART 1 • Multiple Choice

Choose the best answer.

1. There are 40 boys and girls in a club. There are 8 more girls than boys. How many boys are in the club?

   A. 33   C. 16

   B. 29   D. 17

2. There are 30 tables set up for a wedding. The tables have 190 chairs in all. Some of the tables have 6 chairs, and the rest have 8 chairs. How many tables have 6 chairs?

   F. 25   H. 15

   G. 20   I. 5

3. Avocados cost $1 each and papayas cost $3 each. Chaz bought a dozen avocados and papayas for $20. How many avocados did Chaz buy?

   A. 8   C. 6

   B. 7   D. 4

4. Ami is 8 years older than her sister, Ceanne. In four years, Ami will be twice as old as Ceanne. What is the sum of their ages now?

   F. 10   H. 14

   G. 12   I. 16

5. Trisha sends letters and postcards to her friends. She needs 37 cents in postage for each letter and 21 cents for each postcard. She spends $3.11 in postage. How many letters and postcards did Trisha send in all?

   A. 10   C. 12

   B. 11   D. 13

PART 2 • Short Response

Record your answers in the space provided.

6. Franco sees 8 camels and counts 14 humps in all. Some of the camels have two humps and the rest have only one hump. How many camels with two humps did Franco see?

   _____

7. Lemons cost 20 cents each and limes cost 25 cents each. Pia spent $4.40 for twenty lemons and limes. How many lemons did Pia buy?

   _____

PART 3 • Extended Response

Record your answer in the space provided.

8. One hundred packages of socks were sold at the Sock Shoppe last week. Women's socks come in packages of 4 pairs and men's come in packages of 6 pairs. In all, 470 pairs of socks were sold. Show how you can find the numbers of pairs of women's socks and men's socks that were sold.

   _____

   _____

   _____

Lesson 4.6

PART 1 • Multiple Choice

Choose the best answer.

1. How is $2 \times 2 \times 2 \times 2 \times 2$ rewritten using a base and an exponent?

   A. 2^4 C. 32

   B. 5^2 D. 2^5

2. How is 3^4 written in standard form?

   F. 243   H. 64

   G. 81   I. 34

3. How is 12^0 written in standard form?

   A. 0   C. 12×0

   B. 1   D. 12

4. How is $(0.01)^2$ written in standard form?

   F. 0.02   H. 0.0001

   G. 0.001   I. 0.00001

5. $4^{\blacksquare} = 1,024$
   What number completes the number sentence above correctly?

   A. 4   C. 6

   B. 5   D. 256

PART 2 • Short Response

Record your answers in the space provided.

6. How is 14^3 written in standard form?

   _____

7. $5^{\blacksquare} = 3,125$
   What number completes the number sentence above correctly?

   _____

PART 3 • Extended Response

 Record your answer in the space provided.

8. A type of bacteria reproduces by splitting in two every hour. Suppose you begin with 2 bacteria. Explain how exponents can help you find the number of bacteria there will be 24 hours later.

   _____

   _____

   _____

Lesson 5.1

PART 1 • Multiple Choice

Choose the best answer.

1. What is the missing value?

   $21 \div \square = 7$

   A. 21   C. 3

   B. 7   D. 1

2. What is the missing value?

   $8 \times \square = 72$

   F. 9   H. 7

   G. 8   I. 6

3. What is the missing value?

   $\square \div 6 = 5$

   A. 36   C. 24

   B. 30   D. 11

4. $56 \div 7$

   F. 10   H. 8

   G. 9   I. 7

5. $6\overline{)54}$

   A. 9   C. 7

   B. 8   D. 6

PART 2 • Short Response

Record your answers in the space provided.

6. Paula earned $64 working during the week and $8 working over the weekend. How many times as great were her earnings for the week than for the weekend?

   _____

7. Show how the set of numbers 4, 6, and 24 represent a fact family.

   _____

   _____

PART 3 • Extended Response

 Record your answer in the space provided.

8. Martin has 72 books that he wants to distribute evenly on 8 shelves. How many books will he have per shelf? Explain which operation you used to solve the problem and what operation you used to check your answer.

   _____

   _____

Lesson 5.2

PART 1 • Multiple Choice

Choose the best answer.

1. Which of the following is the best estimate of 837 ÷ 4?

   A. 20   C. 200

   B. 30   D. 300

2. Which of the following is the best estimate of 14,876 ÷ 5?

   F. 3,000

   G. 2,000

   H. 300

   I. 200

3. Which of the following is the best estimate of 7,331 ÷ 8?

   A. 80   C. 800

   B. 90   D. 900

4. 21,000 ÷ 30

   F. 70

   G. 700

   H. 7,000

   I. 630,000

5. Notecards are sold in boxes of 20. A carton contains 1,200 notecards. How many boxes are in a carton?

   A. 24,000   C. 60

   B. 600   D. 6

PART 2 • Short Response

Record your answers in the space provided.

6. Find the quotient:
   42,000 ÷ 70

   _____

7. Emily earned $48,000 last year. If she worked 200 days in all, how much did Emily earn each day, on average?

   _____

PART 3 • Extended Response

 Record your answer in the space provided.

8. Explain how you can find 81,000,000 ÷ 9,000 using mental math.

   _____

   _____

   _____

Lesson 5.3

PART 1 • Multiple Choice

Choose the best answer.

1. How many are in each group when 45 is divided by 9?

   A. 3   C. 5

   B. 4   D. 6

2. How many are in each group when 56 is divided by 8?

   F. 3   H. 5

   G. 4   I. 7

3. How many are in each group when 80 is divided by 5?

   A. 16   C. 1

   B. 2   D. 0

4. Divide. $6\overline{)42}$

   F. 7

   G. 6

   H. 5

   I. 4

5. Divide. $7\overline{)63}$

   A. 10

   B. 9

   C. 8

   D. 7

PART 2 • Short Response

Record your answers in the space provided.

6. Four children played a game using stickers. There were 48 stickers in all, and each child received the same number of stickers. How many stickers did each child receive?

   _____

7. Justin baked 60 rolls this morning. Rolls are sold in bags with six rolls in each bag. How many bags will he have?

   _____

PART 3 • Extended Response

 Record your answer in the space provided.

8. Thirty-two students must travel to a tournament by car. Each car can fit four students and a driver. Explain how to find the number of cars needed.

   _____

   _____

Lesson 5.4

PART 1 • Multiple Choice

Choose the best answer.

1. Divide.

   $138 \div 7$

   A. 19 R6   C. 18 R2

   B. 19 R5   D. 16 R6

2. $6\overline{)315}$

   F. 50 R5   H. 52 R2

   G. 51 R3   I. 52 R3

3. $8\overline{)4,153}$

   A. 519 R1   C. 510 R7

   B. 514 R1   D. 506 R5

4. $52,047 \div 6$

   F. 8,107 R5   H. 8,674 R3

   G. 8,617 R3   I. 8,774 R1

5. $47,419 \div 3$

   A. 16,546 R1   C. 15,716 R1

   B. 15,806 R1   D. 12,806 R1

PART 2 • Short Response

Record your answers in the space provided.

6. A farmer has 2,985 tomato plants ready to plant. If she divides the plants equally among 9 rows, how many tomato plants will she have left over?

   _____

7. Sophia and her family went on a 108-mile sailing trip off the coast of Florida. They sailed 9 miles each hour. How many hours did they sail?

   _____

PART 3 • Extended Response

Record your answer in the space provided.

8. When dividing any number by 8, what are the only possible remainders? Explain.

   _____

   _____

   _____

©Macmillan/McGraw-Hill

Lesson 5.5

PART 1 • Multiple Choice

Choose the best answer.

1. One hundred guests will attend a banquet. Each table can seat six guests. How many tables are needed?

   A. 14   C. 16

   B. 15   D. 17

2. A roll of nickels contains 40 nickels. Gina has 235 nickels to be placed into rolls. How many will be left over after she finishes rolling her nickels?

   F. 35   H. 15

   G. 25   I. 5

3. Doughnuts are sold in boxes of 12. If Carmela makes 200 doughnuts, how many boxes can she fill?

   A. 15   C. 17

   B. 16   D. 18

4. On a field trip, 185 students and teachers takes buses to a zoological park. If each bus holds 42 people, how many buses are needed?

   F. 3   H. 5

   G. 4   I. 6

5. An album can hold 16 stamps on each page. Russ has 550 stamps. He does not place any stamps on a page until he can fill the page. How many stamps will Russ have left over if he tries to place his 550 stamps on all the pages?

   A. 0   C. 6

   B. 4   D. 10

PART 2 • Short Response

Record your answers in the space provide.

6. A school with 506 students is divided into classes with 22 students in each class. How many classes are there in this school?

   _____

7. Roses are sold in bunches of 12. If a florist has 700 roses, how many bunches can he make?

   _____

PART 3 • Extended Response

Record your answer in the space provided.

8. The Auburn House has a doorman on duty 24 hours a day, 7 days a week. Suppose you had to decide how long a doorman's normal workday should be. Based on the ideas from this lesson, which shift makes the most sense: 7 hours, 8 hours, or 9 hours? Explain your reasoning.

   _____

   _____

Lesson 6.1

PART 1 • Multiple Choice

Choose the best answer.

1. Which of the following is the most reasonable estimate of 792 ÷ 18?

   A. 30 C. 60

   B. 40 D. 70

2. Which of the following is the most reasonable estimate of 4,125 ÷ 82?

   F. 30 H. 50

   G. 40 I. 60

3. Which of the following is the most reasonable estimate of 83,625 ÷ 425?

   A. 2,000 C. 200

   B. 1,000 D. 100

4. There are about 28 grams in 1 ounce. A package of flour is labeled "2,300 grams." About how many ounces does this package of flour weigh?

   F. 80 H. 800

   G. 100 I. 1,000

5. Lipsha earns $44,562 per year. There are 52 weeks in a year. Which of the following is the most reasonable estimate of Lipsha's weekly earnings?

   A. $900

   B. $1,000

   C. $9,000

   D. $10,000

PART 2 • Short Response

Record your answers in the space provided.

6. Ed estimated 795 ÷ 38 by first rounding the dividend and divisor to the nearest ten. What was Ed's estimate of the quotient?

   _____

7. Wendy estimated 56,047 ÷ 741 by first rounding the dividend and divisor to the nearest hundred. What was Wendy's estimate of the quotient?

   _____

PART 3 • Extended Response

Record your answer in the space provided.

8. To estimate 2,785 ÷ 42, Amos found the quotient of 2,800 ÷ 40. Examine the method Amos used. Explain how you can tell that the actual quotient must be less than the estimate Amos found.

   _____

   _____

   _____

Lesson 6.2

PART 1 • Multiple Choice

Choose the best answer.

1. Which answer choice shows the correct first step for finding 2,318 ÷ 57?

   A. $57\overline{)2318}$ with 5

   B. $57\overline{)2318}$ with 48

   C. $57\overline{)2318}$ with 4

   D. $57\overline{)2318}$ with 40

2. Divide. $32\overline{)4,748}$

   F. 1,411 R14   H. 148 R12

   G. 148 R22   I. 148

3. 54,003 ÷ 83

   A. 65 R53   C. 650 R53

   B. 65 R530   D. 651 R70

4. Ms. Elsen has to record 3,287 scores. She can record 150 scores in 1 hour. How long will it take Ms. Elsen to record these scores?

   F. between 2 and 3 hours

   G. between 3 and 4 hours

   H. between 20 and 21 hours

   I. between 21 and 22 hours

5. 32,652 ÷ 62

   A. 526 R40   C. 542 R58

   B. 542 R48   D. 544 R24

PART 2 • Short Response

Record your answers in the space provided.

6. The Athena Theater has 48 equal-sized rows of seats. There are 1,296 seats in the theater. How many seats are there in each row?

   _____

7. Yuri needs to ship an order of 7,200 ornaments to a store. Each large carton can hold 64 ornaments. Yuri fills as many cartons as he can with 64 ornaments, and then places any ornaments left over in a smaller carton. How many cartons in all does Yuri need to use? How many ornaments will there be in the smaller carton?

   _____

PART 3 • Extended Response

Record your answer in the space provided.

8. When Ariel divided a number by 63, the answer was 35 R13. Explain how you can find Ariel's original number.

   _____

   _____

Lesson 6.3

PART 1 • Multiple Choice

Choose the best answer.

1. Divide.

   $1{,}280 \div 40$

   A. 320   C. 32

   B. 311   D. 31

2. Divide.

   $1{,}524 \div 16$

   F. 91 R8   H. 97 R2

   G. 95 R4   I. 97 R12

3. Divide.

   $42\overline{)5{,}984}$

   A. 118 R32   C. 142 R2

   B. 118 R28   D. 142 R20

4. Divide.

   $20\overline{)41{,}580}$

   F. 2,079   H. 2,709

   G. 2,129   I. 2,790

5. Divide.

   $18\overline{)75{,}583}$

   A. 4,113 R4   C. 4,201 R3

   B. 4,199 R1   D. 4,210 R3

PART 2 • Short Response

Record your answers in the space provided.

6. A grocer ordered 3,312 cans of soup. Each case holds 24 cans. How many cases did he order?

   _____

7. If the quotient of a problem is 82 R17, can the divisor be 15? Explain.

   _____

PART 3 • Extended Response

Record your answer in the space provided.

8. Explain how to find the quotient of $24{,}160 \div 40$ using mental math.

   _____

   _____

   _____

   _____

Lesson 6.4

PART 1 • Multiple Choice

Choose the best answer.

1. Divide. $8\overline{)20.8}$

   A. 0.26   C. 2.6

   B. 2.06   D. 26

2. Divide. $5\overline{)232}$

   F. 46.1   H. 46.3

   G. 46.4   I. 46.5

3. $8.02 \div 100,000$

   A. 0.0802   C. 0.000802

   B. 0.00802   D. 0.0000802

4. Carol used 11 gallons of gas to drive 320.6 miles. To the nearest tenth of a mile, how many miles per gallon did Carol get?

   F. 30.0   H. 29.1

   G. 29.2   I. 29.0

5. A box of 24 crackers weighs 6.4 ounces. How much does one cracker weigh, rounded to the hundredth place?

   A. 0.27 ounces

   B. 0.267 ounces

   C. 0.266 ounces

   D. 0.26 ounces

PART 2 • Short Response

Record your answers in the space provided.

6. What is $563 \div 9$, rounded to the nearest tenth?

   _____

7. A 32-ounce bottle of olive oil costs $4.89. What is the cost of one ounce of this olive oil? Round your answer to the nearest penny.

   _____

PART 3 • Extended Response

 Record your answer in the space provided.

8. How is dividing decimals by whole numbers similar to dividing whole numbers by whole numbers? How is it different?

   _____

   _____

   _____

Lesson 6.5

PART 1 • Multiple Choice

Choose the best answer.

1. Mr. Renny spent $32 for movie tickets. His adult ticket cost $8. The rest of his group were children. Their tickets cost $4 each. How many children were in Mr. Renny's group?

   A. 6   C. 4

   B. 5   D. 3

2. There are 200 seats in an auditorium. Some rows contain 8 seats and 12 rows contain 10 seats. How many rows contain 8 seats?

   F. 8   H. 12

   G. 10   I. 25

3. A television program runs from 9:00 P.M. to 9:30 P.M. There are 16 commercials of equal length during this half hour. The program itself runs for 22 minutes. How long is each commercial?

   A. 8 minutes   C. 0.8 minutes

   B. 1.9 minutes   D. 0.5 minutes

4. An elevator can hold 1,000 pounds. A man who weighs 180 pounds wants to load boxes into the elevator. Each box weighs 40 pounds. How many boxes can this man load on the elevator?

   F. 29   H. 21

   G. 25   I. 20

PART 2 • Short Response

Record your answers in the space provided.

5. Baxter was thinking of a number. He added 10 to his number. Then he divided the sum by 6. Then he subtracted 7 from the quotient. The number he then had was 1. What was Baxter's original number?

   _____

6. An overnight delivery service charged Lana $27 to deliver a package. This service charges $12 plus $3 per pound for delivery. What was the weight of Lana's package?

   _____

PART 3 • Extended Response

Record your answer in the space provided.

7. On March 1, the price of rose bushes tripled. On April 1, the price dropped by $0.75 per bush. On May 1, the price was cut in half. On June 1, the price increased $0.25 to $1.75 per bush. Show the steps you can use to find the price of a bush the day before March 1.

   _____

   _____

   _____

Lesson 7.1

PART 1 • Multiple Choice

Choose the best answer. Use the tally and frequency chart below for problems 1-8.

Number of Books Read Each Month		
Number of Books	Tally	Frequency
1	II	2
2	HIT III	8
3	HIT HIT IIII	14
4	HIT II	7
5	III	3

1. How many students read only 1 book each month?

   A. 1   C. 3

   B. 2   D. 8

2. How many students read exactly 3 books each month?

   F. 7   H. 14

   G. 8   I. 34

3. How many students read more than 3 books each month?

   A. 14   C. 10

   B. 11   D. 8

4. How many students read less than 5 books each month?

   F. 4   H. 21

   G. 15   I. 31

5. How many students read only 1 or 2 books each month?

   A. 2   C. 8

   B. 3   D. 10

PART 2 • Short Response

Record your answers in the space provided.

6. How many students were surveyed?

   _____

7. How can you find the total number of books the students surveyed read?

   _____

   _____

Part 3 • Extended Response

Record your answers in the space provided.

8. Create a line plot using the data from the tally and frequency chart above.

   _____

   _____

©Macmillan/McGraw-Hill

Lesson 7.2

PART 1 • Multiple Choice

Choose the best answer.
Use the data below for problems 1–4.

45, 60, 48, 50, 48, 44, 49, 58

1. What is the median?
   - A. 48
   - B. 48.5
   - C. 49
   - D. 49.5

2. What is the mean?
   - F. 50.25
   - G. 48.5
   - H. 48
   - I. 46.5

3. What is the mode?
   - A. 48
   - B. 48.5
   - C. 49
   - D. 49.5

4. What is the range?
   - F. 13
   - G. 14
   - H. 15
   - I. 16

5. On a national exam, 3 students scored 475, 7 students scored 525, 5 students scored 675, and 5 students scored 650. What is their mean score?
   - A. 586.25
   - B. 565.1
   - C. 575.25
   - D. 650.3

PART 2 • Short Response

Record your answers in the space provided.

6. A store has five clerks. One earns $12 per hour, one earns $15 per hour and three earn $18 per hour. What is the mean hourly wage for these five clerks?

   _____

7. The ages of the 12 players on a basketball team are shown below. What are the median and mode ages on this team?

   22, 32, 30, 26, 25, 30, 31, 25, 25, 28, 29, 21

   _____

PART 3 • Extended Response

Record your answer in the space provided.

8. Four students sell magazine subscriptions to raise money for their school's sports program. Their median number of subscriptions sold was 10 and their mean number of subscriptions sold was 25. How can the two results be so different? Create a set of data that fits the problem.

   _____

   _____

Lesson 7.3

PART 1 • Multiple Choice

Choose the best answer. Use the pictograph below for problems 1–4.

Ice Cream Cones Sold Today

Vanilla	
Chocolate Chip	
Cookie Dough	
Butter Crunch	
Rocky Road	

Each represents 10 ice cream cones

1. Which flavor was most popular?

   A. Chocolate Chip

   B. Cookie Dough

   C. Butter Crunch

   D. Rocky Road

2. How many Rocky Road cones were sold?

   F. 4   H. 30

   G. 25   I. 40

3. How many Butter Crunch cones were sold?

   A. 3   C. 25

   B. 20   D. 30

4. How many more Cookie Dough cones than Vanilla cones were sold?

   F. 48   H. 30

   G. 40   I. 3

PART 2 • Short Response

Record your answers in the space provided.

5. Scott made a pictograph to show the attendance at each performance of a concert. He used a drawing of a piano to represent 25 people. If 225 people came one night, how many symbols did Scott draw?

_____

6. In a pictograph, each symbol stands for 20 votes. Suppose one row of the pictograph contains 6 symbols, plus one half of a symbol. Explain how you can tell what number of votes this row represents.

_____

PART 3 • Extended Response

THINK
SOLVE
EXPLAIN

Record your answer in the space provided.

7. Make a pictograph to represent the data shown in the table below.

Sunny Days in Abbleton Last Year			
February	March	April	May
18	12	10	20

_____

_____

Lesson 7.4

PART 1 • Multiple Choice

Choose the best answer.
160 students from the Harris School were surveyed about computers at home. Use the graph for problems 1–4.

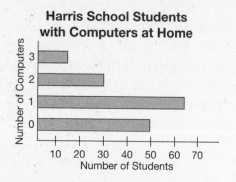

Harris School Students with Computers at Home

1. How many students have no computer at home?

   A. 16   C. 50

   B. 30   D. 64

2. What is the mode number for the students surveyed who have computers at home?

   F. 0   H. 2

   G. 1   I. 3

3. What is the range for the number of computers that students have at home?

   A. 3   C. 48

   B. 4   D. 64

4. What is the median number of computers that students have at home?

   F. 1   H. 1.5

   G. 1.25   I. 1.75

PART 2 • Short Response

Record your answers in the space provided. Use the bar graph for problems 5–6.

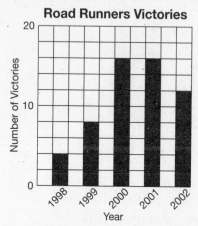

Road Runners Victories

5. How many more victories did the Road Runners have in 2000 than in 1998?

   _____

6. What was the Road Runners' mean number of yearly victories in 1998–2002?

   _____

PART 3 • Extended Response

THINK
SOLVE
EXPLAIN

Record your answer in the space provided.

7. Make a double bar graph to present the data shown.

Average Low and High Temperatures in Bellerton		
Month	January	February
Low	10°F	15°F
High	28°F	40°F

   _____

©Macmillan/McGraw-Hill

Lesson 7.5

PART 1 • Multiple Choice

Choose the best answer. Use the coordinate grid for problems 1–2.

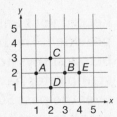

1. What are the coordinates of point E?

   A. (2, 4)   **C.** (4, 2)

   B. (3, 5)   **D.** (5, 3)

2. Which point has coordinates (3, 2)?

   F. Point A   **H.** Point C

   G. Point B   **I.** Point D

Use the line graph for problems 3–5.

Cars Sold at Autorama

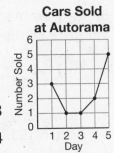

3. How many cars were sold on Day 4?

   A. 1   **C.** 3

   B. 2   **D.** 4

4. On which two days were the same number of cars sold?

   F. Day 1, Day 3

   G. Day 1, Day 4

   H. Day 1, Day 5

   I. Day 2, Day 3

5. Between which two days did car sales increase by the greatest amount?

   A. Day 1, Day 2

   B. Day 2, Day 3

   C. Day 3, Day 4

   D. Day 4, Day 5

PART 2 • Short Response

Record your answers in the space provided.

6. What was the mean number of cars sold each day?

   _____

7. Point X has coordinates (8, 3). Point Y is two units to the left of point X and 5 units above point X. What are the coordinates of point Y?

   _____

PART 3 • Extended Response

Record your answer in the space provided.

8. Describe how you can identify the coordinates of a point on a coordinate grid.

   _____

   _____

   _____

Lesson 7.6

PART 1 • Multiple Choice

Choose the best answer.
Use the graph below for problems 1–2.

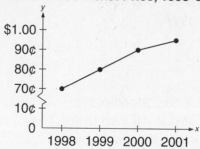

Millerville Bus Ticket Price, 1998–2001

1. In what year was the price of a bus ticket 80¢?

   A. 1998   C. 2000

   B. 1999   D. 2001

2. Suppose the vertical scale did not have a break in it. Instead, the scale went from $0 to $1 in increments of 25¢. How would that change the appearance of the graph?

   F. The price increase over time would appear the same.

   G. The price increase over time would appear greater.

   H. The price increase over time would appear smaller.

   I. There would appear to be no price change over time.

PART 2 • Short Response

Record your answers in the space provided. The bar graph below shows how postage stamp rates have changed over the past 50 years. Use the graph for problems 3–4.

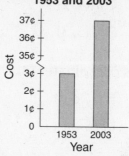

Cost of Mailing a Letter, 1953 and 2003

3. By how many cents did the stamp price increase from 1953 to 2003?

   _____

4. How many times as great as the cost in 1953 is the cost in 2003? Round your answer to the nearest whole number.

   _____

PART 3 • Extended Response

Record your answer in the space provided.

5. Company sales at the Gofferton Company decreased from $100,000 in 2001 to $80,000 in 2002. Make two graphs to shows this data: one that makes the decrease appear small and one that makes the decrease appear great.

Lesson 8.1

PART 1 • Multiple Choice

Choose the best answer. The histogram shows the prices of CDs sold. Use the histogram for problems 1-6.

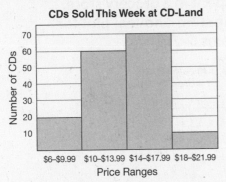

CDs Sold This Week at CD-Land

1. What was the most common price range for CDs sold this week?

   A. $6.00–$9.99

   B. $10.00–$13.99

   C. $14.00–$17.99

   D. $18.00–$21.99

2. How many CDs were sold this week in the price range of $10.00-$13.99?

   F. 70   H. 20

   G. 60   I. 10

3. What was the least common price range for CDs sold this week?

   A. $6.00–$9.99

   B. $10.00–$13.99

   C. $14.00–$17.99

   D. $18.00–$21.99

4. A CD that costs $9.99 is marked as $19.99 by mistake. To correct this, how many bars must be changed?

   F. 4   H. 2

   G. 3   I. 1

PART 2 • Short Response

Record your answers in the space provided.

5. How many CDs were sold in all this week at CD-Land?

   _____

6. How many more CDs were sold for $10.00 or more than for less than $10.00?

   _____

PART 3 • Extended Response

Record your answer in the space provided.

7. The table shows the ages of the people at the Faraway playground one afternoon. Make a histogram for the data in the table.

Age	Number
0–14	18
15–29	24
30–44	6
45–59	10

Lesson 8.2

PART 1 • Multiple Choice

Choose the best answer.
The stem-and-leaf plot shows the ages of the 28 people who work at a company. Use the stem-and-leaf plot for problems 1–5.

**Age of Workers
at the Puretone Company**

stems	leaves
6	0 2 6
5	1 1 8 8
4	0 0 4 5 5 5 9
3	0 1 1 3 3 4 6 8
2	3 4 5 6 7 9

1. What is the range of ages?

   A. 66 years   C. 37 years

   B. 43 years   D. 31 years

2. What is the mode age?

   F. 45   H. 39

   G. 40.5   I. 38

3. What is the median age?

   A. 45   C. 39

   B. 40.5   D. 38

4. What is the mean age?

   F. 45   H. 39

   G. 40.5   I. 38

5. The oldest worker retires. Which of the following would NOT change?

   A. mean   C. mode

   B. median   D. range

PART 2 • Short Response

Record your answers in the space provided. A survey of shoppers timed how long each spent in a grocery store. Use the stem-and-leaf plot for problems 6–7.

**Minutes Spent
in a Grocery Store**

stems	leaves
3	4 4 6 6 9 9
2	5 5 5 6 8 8 8
1	1 5 7 7 9
0	5 8 9

6. How many shoppers were surveyed?

   _____

7. What was the greatest number of minutes spent in the grocery store?

   _____

PART 3 • Extended Response

 Record your answer in the space provided.

8. Willie Mays's season home run totals were 20, 4, 41, 51, 36, 35, 29, 34, 29, 40, 49, 38, 47, 52, 37, 22, 23, 13, 28, 18, 8, and 6. Make a stem-and-leaf plot to display his season home run totals. Then find his mean and median season home run totals.

   _____

Lesson 8.3

PART 1 • Multiple Choice

Choose the best answer.

1. A scientist wants to show how the temperature changes over a day. Which type of graph best displays this data?

   A. double bar graph
   B. histogram
   C. line graph
   D. pictograph

2. A student wants to compare how much time men and women spend watching different sports on television. Which type of graph best displays this data?

   F. double bar graph
   G. histogram
   H. line graph
   I. stem-and-leaf plot

3. Which of the following is best displayed using a stem-and-leaf plot?

   A. the ages of 30 players on a football team
   B. the salaries of 500 workers at a company
   C. the low temperatures over the course of 10 days
   D. the scores of 500 students on a state exam

4. Todd wants to display the distances of the planets from the sun. Which type of graph best displays this data?

   F. bar graph
   G. double bar graph
   H. histogram
   I. line graph

PART 2 • Short Response

Record your answers in the space provided.

5. Alice used a stem-and-leaf plot to display the scores of 50 students on an exam. The mean score was 84, the median score was 80, the highest score was 98, and the lowest score was 58. How many stems did Alice's graph have?

   _____

6. Which type of data display is best to use when you want to show changes over time?

   _____

PART 3 • Extended Response

 Record your answer in the space provided.

7. Describe a situation where you would choose to use a double bar graph instead of a bar graph.

   _____

   _____

Lesson 8.4

Part 1 • Multiple Choice

Choose the best answer.

1. In a class of 28 students, you pick 5 names at random. You want to find out how many students in the class play sports. What is the sample?

   A. 28 students

   B. 5 students

   C. students that play sports

   D. students that do not play sports

2. You pick 30 students at random from the list of students in the school. You want to see who is interested in an intramural sports program. What is the population?

   F. 30 students

   G. students in the school

   H. students in the district

   I. students in the town

3. A researcher wants to find out which car color is the most popular in your town. Which sample is random?

   A. a survey of the first half of the phone book

   B. a survey of every shopper at one local car dealer

   C. a survey of every twentieth person in the phone book

   D. a survey of every tenth person over 30 years old

Part 2 • Short Response

Record your answers in the space provided.

4. A survey of 235 people revealed that 196 of them do not eat breakfast. How many of the people surveyed eat breakfast?

   _____

5. If you want to conduct a survey for a store to find out the favorite soda among shoppers, how can you find a representative sample?

   _____

Part 3 • Extended Response

 Record your answer in the space provided.

6. A researcher wants to find the most popular brand of shoes at your school. Explain how the survey can be conducted so that it is a random sample and representative of the population. Justify your answer.

   _____

   _____

Lesson 9.1

PART 1 • Multiple Choice

Choose the best answer.

1. By which number is 4,203 divisible?
   - A. 2
   - C. 6
   - B. 5
   - D. 9

2. By which number is 10,101 divisible?
   - F. 3
   - H. 9
   - G. 5
   - I. 10

3. By which number is 852 NOT divisible?
   - A. 5
   - C. 3
   - B. 4
   - D. 2

4. By which number is 81,780 NOT divisible?
   - F. 10
   - H. 6
   - G. 9
   - I. 5

5. Kyle has to set up 102 folding chairs in an auditorium. How many rows of chairs should Kyle make, if he wants all rows to have the same number of chairs?
   - A. 5
   - B. 6
   - C. 9
   - D. 10

PART 2 • Short Response

Record your answers in the space provided.

6. What is the least number greater than 50 that is divisible by 4, 5, and 6?

   _____

7. By which numbers less than 10 is 6,548 divisible?

   _____

PART 3 • Extended Response

 Record your answer in the space provided.

8. You know that a number is divisible by 6 if it is divisible by both 2 and 3. Claire claims that a number is divisible by 12 if it is divisible by both 2 and 6. Do you agree or disagree with Claire? Explain your thinking.

   _____

   _____

   _____

Lesson 9.2

PART 1 • Multiple Choice

Choose the best answer.

1. Which number is prime?

   A. 8   C. 10

   B. 9   D. 11

2. Which number is composite?

   F. 29   H. 21

   G. 23   I. 19

3. What is the prime factorization of 20?

   A. 1×20

   B. 4×5

   C. 2×10

   D. $2^2 \times 5$

4. What is the prime factorization of 72?

   F. $2^2 \times 3^2$

   G. $2^2 \times 3^3$

   H. $2^3 \times 3^2$

   I. $2^3 \times 3^3$

5. What is the prime factorization of 250?

   A. $2^3 \times 5^2$

   B. 2×5^3

   C. $2 \times 3 \times 5^2$

   D. $2^2 \times 3 \times 5^2$

PART 2 • Short Response

Record your answers in the space provided.

6. What is the greatest two-digit prime number?

   _____

7. Complete the sentence: A composite number must be divisible by at least _____ whole numbers.

   _____

PART 3 • Extended Response

Record your answer in the space provided.

8. Explain why any three-digit number that ends in 0, 2, 4, 5, 6, or 8 cannot be prime.

   _____

   _____

   _____

Lesson 9.3

PART 1 • Multiple Choice

Choose the best answer.

1. What is the greatest common factor of 12 and 36?

   A. 4   C. 12

   B. 6   D. 36

2. What is the greatest common factor of 15 and 44?

   F. 1   H. 3

   G. 2   I. 5

3. What is the greatest common factor of 24 and 54?

   A. 8   C. 4

   B. 6   D. 3

4. How many common factors do 30 and 35 have?

   F. 1

   H. 3

   G. 2

   I. 4

5. What is the greatest common factor of 48 and 49?

   A. 4

   C. 2

   B. 3

   D. 1

PART 2 • Short Response

Record your answers in the space provided.

6. What is the GCF of 60, 80 and 105?

   _____

7. How many common factors do 50 and 75 have?

   _____

PART 3 • Extended Response

Record your answer in the space provided.

8. A box of chocolate cookies contains 24 cookies. A box of lemon cookies contains 40 cookies. A group of children shared the cookies in each box. Each child got an equal number of cookies from each box. What is the greatest number of children there could be in this group? Explain your reasoning.

   _____

   _____

   _____

Lesson 9.4

PART 1 • Multiple Choice

Choose the best answer.

1. What fraction is shown by the shaded part of the circle?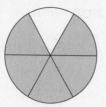

    A. $\frac{5}{6}$ C. $\frac{6}{5}$

    B. $\frac{5}{1}$ D. $\frac{6}{1}$

2. In which model is $\frac{1}{3}$ shaded?

    F.   H.

    G.   I.

3. A pizza has 8 slices. Charlene ate 2 slices and Rafael ate 3 slices. What fraction of the pizza did they eat?

    A. $\frac{3}{8}$ C. $\frac{5}{8}$

    B. $\frac{5}{3}$ D. $\frac{8}{5}$

4. Thirteen boys and 11 girls are in a class. What fraction of the class is girls?

    F. $\frac{11}{13}$ H. $\frac{11}{24}$

    G. $\frac{13}{11}$ I. $\frac{24}{11}$

5. Theo brought 36 apples to school. His class ate 29 of the apples. What fraction of the apples remained?

    A. $\frac{7}{36}$ C. $\frac{36}{7}$

    B. $\frac{17}{36}$ D. $\frac{36}{17}$

PART 2 • Short Response

Record your answers in the space provided.

6. A large cake was cut into 20 slices, as modeled below. The Kullens ate $\frac{1}{4}$ of the cake for dessert. How many slices did the Kullens eat?

    _____

7. Leah attends school every day from Monday through Friday. She baby-sits every day from Thursday through Sunday. What fraction of the days in a week does Leah both attend school and baby-sit?

    _____

PART 3 • Extended Response

THINK SOLVE EXPLAIN Record your answer in the space provided.

8. Draw a square and use it to model the fraction $\frac{13}{16}$. Explain your method.

    _____

    _____

    _____

Lesson 9.5

PART 1 • Multiple Choice

Choose the best answer.

1. Which fraction is equivalent to $\frac{3}{5}$?

   A. $\frac{2}{4}$ C. $\frac{6}{10}$

   B. $\frac{4}{6}$ D. $\frac{8}{10}$

2. Which fraction is equivalent to $\frac{9}{12}$?

   F. $\frac{12}{9}$ H. $\frac{3}{6}$

   G. $\frac{6}{9}$ I. $\frac{3}{4}$

3. Which fraction is NOT equivalent to $\frac{2}{4}$?

   A. $\frac{1}{2}$ C. $\frac{5}{10}$

   B. $\frac{4}{6}$ D. $\frac{8}{16}$

4. Choose the missing number that completes the number sentence.

   $$\frac{7}{8} = \frac{d}{16}$$

   F. 8   H. 14

   G. 12   I. 15

5. Choose the missing number that completes the number sentence.

   $$\frac{4}{12} = \frac{1}{a}$$

   A. 3   C. 6

   B. 4   D. 9

PART 2 • Short Response

Record your answers in the space provided.

6. A fraction is equivalent to $\frac{5}{8}$ and has a numerator of 20. What is the fraction's denominator?

   _____

7. Write two fractions that are equivalent to $\frac{10}{12}$.

   _____

PART 3 • Extended Response

Record your answer in the space provided.

8. Ann and Ben played softball on 24 of the 30 days in June. Ann said, "We played softball on $\frac{12}{15}$ of the days." Ben replied, "No, we played softball on $\frac{4}{5}$ of the days." Is either correct? Explain your thinking.

   _____

   _____

   _____

Lesson 9.6

PART 1 • Multiple Choice

Choose the best answer.

1. What is $\frac{20}{25}$ written in simplest form?

   A. $\frac{4}{5}$ C. $\frac{2}{5}$

   B. $\frac{2}{3}$ D. $\frac{1}{5}$

2. What is $\frac{12}{48}$ written in simplest form?

   F. $\frac{1}{36}$ H. $\frac{1}{4}$

   G. $\frac{1}{11}$ I. $\frac{2}{3}$

3. Which fraction is written in simplest form?

   A. $\frac{3}{12}$ C. $\frac{8}{14}$

   B. $\frac{15}{30}$ D. $\frac{2}{3}$

4. Which fraction is written in simplest form?

   F. $\frac{2}{8}$ H. $\frac{4}{8}$

   G. $\frac{3}{8}$ I. $\frac{6}{8}$

5. Choose the number that completes the number sentence.

   $$\frac{48}{56} = \frac{\square}{7}$$

   A. 12   C. 6

   B. 8   D. 5

PART 2 • Short Response

Record your answers in the space provided.

6. Jared read 28 out of 60 pages in his book. If the number of pages Jared read is expressed as a fraction in simplest form, what will the denominator be?

   _____

7. Lisa has 18 throw pillows for her daybed. Twelve of them are red and the rest are green. Explain how to find the fraction, in simplest form, of the throw pillows that are green?

   _____

PART 3 • Extended Response

Record your answer in the space provided.

8. A garden contains 24 tomato plants, 11 pepper plants, 15 zucchini plants, and 10 basil plants. What fraction of the garden is each type of plant, expressed in simplest form? Which fraction is already in simplest form? Explain.

   _____

   _____

©Macmillan/McGraw-Hill

Lesson 9.7

PART 1 • Multiple Choice

Choose the best answer.
Use the table below for problems 1–6.

Meadville Recreation Center
Craft Courses

Courses	Sign-ups	FEES
Pottery	15	**Residents**
Wood Carving	12	$25 per course
Jewelry Making	18	Maximum $100
Flower Arranging	10	**Nonresidents**
Glassblowing	20	$30 per course
		Maximum $125

1. What fraction of people signed up for glassblowing?

    A. $\frac{4}{15}$ C. $\frac{20}{65}$

    B. $\frac{20}{55}$ D. There is not enough information.

2. How much did residents pay in all to take flower arranging?

    F. $125   H. $250

    G. $150   I. There is not enough information.

3. You want to know how much Maria spent for her courses. Which below is *extra information* that you do not need to solve this problem?

    A. Maria is a resident.

    B. There are 18 students in one of Maria's classes.

    C. Maria is taking more than two courses.

    D. Maria is taking fewer than four courses.

PART 2 • Short Response

Record your answers in the space provided.

4. Jackson is a non-resident who signed up for 4 courses. If he decides to take a fifth course, how much more will the course cost him?

5. In one of the classes, exactly $\frac{7}{9}$ of the students are residents. Which class is this?

PART 3 • Extended Response

Record your answer in the space provided.

6. Tina wants to take two courses and has $55 to spend. Does she have enough money? Explain your reasoning.

Lesson 10.1

PART 1 • Multiple Choice

Choose the best answer.

1. What is the LCM of 4 and 16?

   A. 4   C. 32

   B. 16   D. 64

2. What is the LCD of $\frac{5}{6}$ and $\frac{2}{9}$?

   F. 9   H. 18

   G. 10   I. 54

3. How are $\frac{2}{3}$ and $\frac{3}{4}$ rewritten as equivalent fractions using their LCD?

   A. $\frac{8}{12}$ and $\frac{9}{12}$

   B. $\frac{2}{12}$ and $\frac{3}{12}$

   C. $\frac{4}{6}$ and $\frac{5}{6}$

   D. $\frac{2}{6}$ and $\frac{3}{6}$

4. Which of the following is NOT a common multiple of 6 and 8?

   F. 96   H. 48

   G. 72   I. 36

5. Hot dogs are sold in packages of 10. Hot dog rolls are sold in packages of 8. Stan wants to buy the same number of each. What is the least number of hot dogs Stan can buy?

   A. 20

   B. 40

   C. 60

   D. 80

PART 2 • Short Response

Record your answers in the space provided.

6. What is the LCM of 4, 5, and 6?

   _____

7. Jeannie did her laundry and mopped her kitchen floor on March 1. She does her laundry every 3 days and mops her kitchen floor every 5 days. On what date will Jeannie next do her laundry and mop her kitchen floor on the same day?

   _____

PART 3 • Extended Response

Record your answer in the space provided.

8. Keyshawn is planning his soccer practice and his piano practice for the next 30 days. He plans to practice piano every third day and to practice soccer every other day. On how many days will he practice both soccer and piano? Explain your method.

   _____

   _____

   _____

©Macmillan/McGraw-Hill

Lesson 10.2

PART 1 • Multiple Choice

Choose the best answer.

1. Which pair of fractions are equivalent?

    A. $\frac{2}{3}$ and $\frac{5}{9}$

    B. $\frac{3}{4}$ and $\frac{9}{12}$

    C. $\frac{1}{2}$ and $\frac{6}{10}$

    D. $\frac{1}{4}$ and $\frac{4}{8}$

2. Select the statement that is true.

    F. $\frac{1}{3}$ $\frac{1}{2}$ H. $\frac{1}{5}$ $\frac{1}{6}$

    G. $\frac{1}{4}$ $\frac{1}{3}$ I. $\frac{1}{7}$ $\frac{1}{5}$

3. Select the statement that is true.

    A. $\frac{3}{8}$ $\frac{1}{4}$ C. $\frac{2}{3}$ $\frac{5}{12}$

    B. $\frac{3}{4}$ $\frac{5}{16}$ D. $\frac{4}{5}$ $\frac{7}{8}$

4. Which set of numbers is ordered from least to greatest?

    F. $\frac{1}{8}, \frac{1}{5}, \frac{1}{2}$ H. $\frac{1}{5}, \frac{1}{2}, \frac{1}{8}$

    G. $\frac{1}{2}, \frac{1}{5}, \frac{1}{8}$ I. $\frac{1}{8}, \frac{1}{2}, \frac{1}{5}$

5. Which set of numbers is ordered from least to greatest?

    A. $\frac{1}{3}, \frac{3}{4}, \frac{3}{8}$ C. $\frac{3}{8}, \frac{1}{3}, \frac{3}{4}$

    B. $\frac{1}{3}, \frac{3}{8}, \frac{3}{4}$ D. $\frac{3}{8}, \frac{3}{4}, \frac{1}{3}$

PART 2 • Short Response

Record your answers in the space provided.

6. Ted had $\frac{3}{5}$ of his work completed and Gregg had $\frac{5}{8}$ of his work completed. To compare who has more of their work completed, find the LCD for the fractions.

    _____

7. In a group of marbles, $\frac{2}{9}$ are blue and $\frac{3}{4}$ are green. Are there more blue marbles or green marbles? Explain.

    _____

    _____

PART 3 • Extended Response

Record your answer in the space provided.

8. A student claims that he read $\frac{1}{2}$ of his book on Monday and $\frac{5}{8}$ on Tuesday. He did not read any pages twice. Is this possible? Explain.

    _____

    _____

Lesson 10.3

PART 1 • Multiple Choice

Choose the best answer.

1. How is 0.25 written as a fraction in simplest form?

   A. $\frac{25}{100}$ C. $\frac{5}{2}$

   B. $\frac{5}{20}$ D. $\frac{1}{4}$

2. How is $\frac{4}{5}$ written as a decimal?

   F. 4.5   H. 0.8

   G. 0.45   I. 0.08

3. Which pair of numbers is equivalent?

   A. $\frac{1}{2}$ and 0.12

   B. $\frac{3}{8}$ and 0.35

   C. $\frac{3}{4}$ and 0.75

   D. $\frac{7}{10}$ and 0.07

4. Which pair of numbers is NOT equivalent?

   F. $\frac{2}{3}$ and 0.68

   G. $\frac{2}{5}$ and 0.4

   H. $\frac{2}{8}$ and 0.25

   I. $\frac{2}{10}$ and 0.2

5. The thickness of a sheet of plywood is $\frac{5}{8}$ inches. How is this written as a decimal?

   A. 0.5 inches

   B. 0.58 inches

   C. 0.625 inches

   D. 0.75 inches

PART 2 • Short Response

Record your answer in the space provided.

6. Of the 25 students involved in student government, 13 are fifth graders. Write the fraction of students who are fifth graders as a decimal.

   _____

PART 3 • Extended Response

Record your answer in the space provided.

7. Jay was sent to the store by his mother. Her list is shown below.

   $\frac{1}{4}$ pound of American cheese

   $\frac{1}{2}$ pound of Ham

   $\frac{3}{4}$ pound of Turkey

   When Jay orders the items shown on the list, the deli scale displays weights using decimals. Explain how Jay can check to get the right amounts.

   _____

   _____

Lesson 10.4

PART 1 • Multiple Choice

Choose the best answer.
A video store kept track of the number of video games it rented each week. Use the table below for problems 1–7.

Number of Video Games Rented

Week	Number	Week	Number	Week	Number
1	39	6	48	11	38
2	47	7	52	12	49
3	52	8	54	13	43
4	53	9	60	14	38
5	37	10	52	15	36

1. In what fraction of the weeks was the number of games rented between 31 and 40?

   A. $\frac{1}{2}$　　C. $\frac{3}{5}$

   B. $\frac{1}{3}$　　D. $\frac{4}{15}$

2. In what fraction of the weeks was the number of games rented greater than 50?

   F. $\frac{1}{3}$　　H. $\frac{1}{5}$

   G. $\frac{2}{3}$　　I. $\frac{2}{5}$

3. In Week 16, 33 games were rented. Now, in what fraction of the weeks was the number of games rented between 31 and 40?

   A. $\frac{1}{3}$　　C. $\frac{3}{8}$

   B. $\frac{2}{5}$　　D. $\frac{5}{16}$

4. What was the mode number of weekly rentals?

   F. 38　　H. 50

   G. 48　　I. 52

PART 2 • Short Response

Record your answers in the space provided.

5. Find the mean number of weekly rentals. Round your answer to one decimal place.

   _____

6. Find the median number of weekly rentals.

   _____

PART 3 • Extended Response

Record your answer in the space provided.

7. Instead of making a table, Ross made a stem-and-leaf plot for the video game data. Make your own stem-and-leaf plot here.

Lesson 10.5

PART 1 • Multiple Choice

Choose the best answer.

1. How is $4\frac{3}{4}$ written as a decimal?

   A. 1.75   C. 4.72

   B. 4.34   D. 4.75

2. How is 6.08 written as a mixed number in simplest form?

   F. $6\frac{4}{5}$ H. $6\frac{2}{25}$

   G. $6\frac{1}{12}$ I. $6\frac{1}{125}$

3. How is $\frac{21}{9}$ written as a mixed number in simplest form?

   A. $2\frac{1}{3}$ C. $2\frac{1}{9}$

   B. $2\frac{1}{6}$ D. $2\frac{2}{9}$

4. How is $6\frac{5}{6}$ written as an improper fraction?

   F. $\frac{36}{6}$ H. $\frac{65}{6}$

   G. $\frac{41}{6}$ I. $\frac{36}{5}$

5. A space mission lasts 10 days and 3 hours. How is this amount of time expressed using mixed numbers?

   A. $10\frac{3}{10}$ days

   B. $10\frac{1}{8}$ days

   C. $10\frac{1}{4}$ days

   D. $10\frac{3}{8}$ days

PART 2 • Short Response

Record your answers in the space provided.

6. A poster is $21\frac{1}{4}$ inches wide. How is this number written as a decimal?

   _____

7. A piece of fabric 94 inches long is divided into 6 pieces of equal length. How long is each piece? Write your answer as a mixed number in simplest form.

   _____

PART 3 • Extended Response

Record your answer in the space provided.

8. A piece of land that is 250 feet wide is going to be divided into 4 lots of equal width. Show three different ways that the width of the smaller lots can be expressed. Then explain which choice you would use to describe the width to a friend and why.

   _____

   _____

   _____

Lesson 10.6

PART 1 • Multiple Choice

Choose the best answer.

1. Which number is greatest?

   A. $\frac{1}{2}$ C. $\frac{1}{4}$

   B. $\frac{1}{3}$ D. $\frac{2}{5}$

2. Which number is least?

   F. $7\frac{2}{3}$ H. $7\frac{7}{9}$

   G. $8\frac{1}{10}$ I. $7\frac{5}{8}$

3. Which set of numbers is ordered from least to greatest?

   A. $5\frac{2}{3}$, 5.9, $5\frac{3}{4}$

   B. $5\frac{2}{3}$, $5\frac{3}{4}$, 5.9

   C. $5\frac{3}{4}$, $5\frac{2}{3}$, 5.9

   D. 5.9, $5\frac{3}{4}$, $5\frac{2}{3}$,

4. Which set of numbers is ordered from greatest to least?

   F. $4\frac{1}{6}$, 4.15, $4\frac{1}{8}$

   G. $4\frac{1}{6}$, $4\frac{1}{8}$, 4.15

   H. $4\frac{1}{8}$, $4\frac{1}{6}$, 4.15

   I. $4\frac{1}{8}$, 4.15, $4\frac{1}{6}$

5. Which number is less than 2.3?

   A. $2\frac{1}{3}$

   B. $2\frac{1}{2}$

   C. $2\frac{2}{9}$

   D. $2\frac{3}{10}$

PART 2 • Short Response

Record your answers in the space provided.

6. Aaron is thinking of a decimal number that has one place to the right of its decimal point. The number is greater than $3\frac{2}{3}$ and less than $3\frac{3}{4}$. What is Aaron's number?

   _____

7. Jasmine is thinking of a mixed number greater than 4.4 and less than 4.5. The denominator of the fraction in the mixed number is 12. What is the mixed number?

   _____

PART 3 • Extended Response

 Record your answer in the space provided.

8. Geena has one wrench whose opening is $\frac{7}{8}$ inch and a larger wrench whose opening is $\frac{15}{16}$ inch. Unfortunately, the bolt she is trying to loosen is too big for the first wrench and too small for the second wrench. How wide is the bolt? Find all the possible answers written to two decimal places. Show your method.

   _____

   _____

Lesson 11.1

PART 1 • Multiple Choice

Choose the best answer.

1. $\frac{5}{12} + \frac{3}{12} =$

   A. $\frac{3}{4}$ C. $\frac{8}{24}$

   B. $\frac{2}{3}$ D. $\frac{1}{16}$

2. $\frac{5}{6} + \frac{5}{6} =$

   F. $\frac{10}{12}$ H. $1\frac{2}{3}$

   G. $1\frac{1}{3}$ I. $1\frac{5}{6}$

3. $\frac{13}{16} + \frac{7}{16} =$

   A. $\frac{20}{23}$ C. $1\frac{3}{16}$

   B. $1\frac{1}{8}$ D. $1\frac{1}{4}$

4. $2\frac{7}{8} + 3\frac{5}{8} =$

   F. $6\frac{1}{2}$ H. $6\frac{1}{4}$

   G. $6\frac{3}{8}$ I. $5\frac{12}{16}$

5. Select the statement that is true.

   A. $\frac{2}{3}$ $\frac{2}{3}$ $\frac{3}{12}$ $\frac{5}{12}$

   B. $\frac{3}{4}$ $\frac{2}{4}$ $\frac{5}{8}$ $\frac{7}{8}$

   C. $\frac{5}{8}$ $\frac{3}{8}$ $\frac{5}{12}$ $\frac{7}{12}$

   D. $\frac{3}{5}$ $\frac{4}{5}$ $\frac{7}{16}$ $\frac{5}{16}$

PART 2 • Short Response

Record your answers in the space provided.

6. Kyle sold $\frac{3}{8}$ of his cookies on Monday and $\frac{4}{8}$ of his cookies on Tuesday. He added the fractions together to find the total fraction of the cookies he sold over the two days. What was Kyle's numerator?

   _____

7. Find the sum of $2\frac{3}{8} + 1\frac{5}{8}$ and create a model to illustrate the answer.

   _____

PART 3 • Extended Response

Record your answer in the space provided.

8. Explain why $\frac{5}{12} + \frac{2}{12} = \frac{7}{12}$ and not $\frac{7}{24}$. Use a group of 12 objects in your explanation.

   _____

   _____

   _____

Lesson 11.2

PART 1 • Multiple Choice

Choose the best answer.

1. In the first week of April, rainfall totaled $1\frac{1}{16}$ inches. In the second week of April, rainfall totaled $1\frac{7}{16}$ inches. How much more rain fell in the second week?

   A. $\frac{3}{8}$ inch   C. $1\frac{3}{8}$ inches

   B. $\frac{1}{2}$ inch   D. $2\frac{1}{2}$ inches

2. A snowstorm in early January dropped $8\frac{1}{2}$ inches on Timber Falls. A snowstorm in late January dropped $5\frac{1}{2}$ inches on Timber Falls. How much snow fell in both storms?

   F. 14 inches   H. 13 inches

   G. $13\frac{1}{2}$ inches   I. 3 inches

3. The table shows the height of Agnes's avocado plant on May 1 and June 1. How much did the plant grow during the month of May?

Date	Plant Height
May 1	$2\frac{1}{4}$ feet
June 1	$2\frac{3}{4}$ feet

   A. 5 feet   C. $\frac{1}{2}$ foot

   B. $4\frac{1}{2}$ feet   D. $\frac{1}{4}$ foot

4. In a 1-mile race, the horse Timely ran the last $\frac{1}{8}$ mile in $22\frac{1}{5}$ seconds. In his next 1-mile race, he ran the last $\frac{1}{8}$ mile in $21\frac{4}{5}$ seconds. How much faster did Timely run the last $\frac{1}{8}$ mile in the second race?

   F. $\frac{7}{8}$ mile   H. $1\frac{2}{5}$ seconds

   G. 44 seconds   I. $\frac{2}{5}$ second

PART 2 • Short Response

Record your answers in the space provided.

5. Mrs. Jansen asked her daughter Christine to buy $2\frac{3}{4}$ pounds of ground meat to make chili. Christine selected a packet of meat labeled 2.86 pounds. How many extra pounds of meat did Christine buy? Write your answer as a decimal.

_____

6. Ferdy ran in three races last month. Their lengths were $1\frac{1}{2}$ miles, $2\frac{1}{2}$ miles, and $\frac{1}{2}$ mile. What was the total distance in Ferdy's races last month?

_____

PART 3 • Extended Response

Record your answer in the space provided.

7. Ellen knows that it is $6\frac{3}{4}$ miles to the town of Reddington where her sister lives. As she is riding her bike there, she passes the sign shown above. Is Ellen more than half of the way to Reddington? Explain your reasoning.

Reddington
3.2 miles

_____

_____

Lesson 11.3

PART 1 • Multiple Choice

Choose the best answer.

1. $\frac{1}{2} + \frac{1}{4} =$

   A. $\frac{1}{6}$ C. $\frac{3}{8}$

   B. $\frac{2}{6}$ D. $\frac{3}{4}$

2. $\frac{3}{10} + \frac{2}{5} =$

   F. $\frac{7}{20}$ H. $\frac{5}{10}$

   G. $\frac{7}{10}$ I. $\frac{5}{15}$

3. $\frac{2}{3} + \frac{1}{6} =$

   A. $\frac{3}{6}$ C. $\frac{3}{9}$

   B. $\frac{5}{6}$ D. $\frac{5}{12}$

4. $\frac{3}{5} + \frac{1}{4} =$

   F. $\frac{4}{9}$ H. $\frac{17}{20}$

   G. $\frac{4}{20}$ I. $\frac{17}{40}$

5. On the first day of Lanier's vacation, $\frac{1}{4}$ inch of rain fell. On the second day, $\frac{3}{8}$ inch of rain fell. How much rain fell in all during these two days?

   A. $\frac{5}{8}$ inch

   B. $\frac{1}{8}$ inch

   C. $\frac{7}{16}$ inch

   D. $\frac{5}{16}$ inch

PART 2 • Short Response

Record your answers in the space provided.

6. How is the sum $\frac{1}{3} + \frac{1}{6}$ written as a decimal?

   _____

7. On Monday, Jayne's stock rose $\frac{3}{8}$ per share. On Tuesday, it rose $\frac{1}{2}$ per share. How much did the stock rise per share in all during these two days? Write your answer as a fraction in simplest form.

   _____

PART 3 • Extended Response

 Record your answer in the space provided.

8. Draw a picture that shows why $\frac{1}{2} + \frac{3}{10} = \frac{4}{5}$.

   _____

   _____

©Macmillan/McGraw-Hill

Lesson 11.4

PART 1 • Multiple Choice

Choose the best answer.

1. $\begin{array}{r} \frac{1}{3} \\ +\frac{3}{4} \\ \hline \end{array}$

    A. $\frac{4}{12}$ C. $\frac{11}{12}$

    B. $\frac{4}{7}$ D. $1\frac{1}{12}$

2. $\frac{4}{5}$ $\frac{3}{10}$

    F. $\frac{7}{15}$ H. $1\frac{1}{15}$

    G. $1\frac{1}{10}$ I. $1\frac{2}{5}$

3. $\frac{9}{16}$ $\frac{7}{8}$

    A. $\frac{16}{24}$ C. $1\frac{7}{16}$

    B. $1\frac{1}{8}$ D. $1\frac{9}{16}$

4. $\frac{1}{4} + \frac{5}{6} + \frac{1}{12} =$

    F. $\frac{7}{22}$ H. $1\frac{1}{6}$

    G. $\frac{7}{12}$ I. $1\frac{3}{12}$

5. A stock rose $\frac{7}{8}$ per share on Wednesday and $\frac{3}{4}$ per share on Thursday. How much did the stock rise over the two days?

    A. $1\frac{5}{8}$ C. $1\frac{1}{4}$

    B. $1\frac{3}{8}$ D. $\frac{10}{12}$

Part 2 Short Response

Record your answers in the space provided.

6. Sabrina read for $\frac{1}{2}$ of an hour on Tuesday and $\frac{7}{8}$ of an hour on Thursday. To find the total time she read, what is the LCD needed to add the fractions together?

    _____

7. Draw a model to show how $\frac{1}{2}$ is changed into an equivalent fraction to add it to $\frac{3}{4}$.

    _____

PART 3 • Extended Response

 Record your answer in the space provided.

8. Victor and Mark work for a printing business and keep track of their time for each job they do. Victor worked $\frac{3}{4}$ of an hour on his first printing job and $\frac{1}{2}$ of an hour on his second printing job. Mark worked $\frac{1}{4}$ of an hour on his first printing job and $\frac{5}{8}$ of an hour on his second printing job. Who worked the greatest amount of time on the two jobs combined? Show your steps.

    _____

Lesson 11.5

PART 1 • Multiple Choice

Choose the best answer.

1. $\begin{array}{r} 2\frac{1}{10} \\ +1\frac{1}{5} \\ \hline \end{array}$

   A. $3\frac{1}{50}$ C. 5

   B. $3\frac{2}{15}$ D. $3\frac{3}{10}$

2. $4\frac{2}{3} + 3\frac{1}{8} =$

   F. $7\frac{3}{11}$ H. $7\frac{11}{24}$

   G. $7\frac{3}{24}$ I. $7\frac{19}{24}$

3. $5\frac{3}{5} + 5\frac{3}{4} =$

   A. $10\frac{3}{9}$ C. $11\frac{7}{20}$

   B. $10\frac{7}{20}$ D. $11\frac{17}{20}$

4. On Thursday, $1\frac{3}{4}$ inches of snow fell. On Friday, $4\frac{1}{2}$ inches of snow fell. How much snow fell in all?

   F. $5\frac{1}{4}$ inches H. $6\frac{1}{4}$ inches

   G. $5\frac{2}{3}$ inches I. $6\frac{2}{3}$ inches

5. A bookcase has shelves that are $18\frac{1}{4}$ inches wide. In addition, the sides of the bookcase are $1\frac{7}{8}$ inches wide on each of its two sides. What is the total width of this bookcase?

   A. 22 inches C. 21 inches

   B. $21\frac{15}{16}$ inches D. $20\frac{1}{4}$ inches

PART 2 • Short Response

Record your answers in the space provided.

6. The mixed numbers $2\frac{3}{8}$, $3\frac{1}{6}$ and $5\frac{2}{3}$ are added together. Then the sum is written in simplest form. What is the denominator of the fractional part of the sum?

   ____

7. Bea worked $37\frac{1}{2}$ hours of regular time last week and $12\frac{3}{4}$ hours of overtime. How many hours did Bea work in all?

   ____

PART 3 • Extended Response

THINK SOLVE EXPLAIN

Record your answer in the space provided.

8. Find the sum $1\frac{2}{3} + 1\frac{3}{4}$. Then draw a picture to show why the sum is greater than 3.

   ____

   ____

   ____

©Macmillan/McGraw-Hill

Lesson 11.6

PART 1 • Multiple Choice

Choose the best answer.

1. Which property is used to rewrite the sum?

   $$4\tfrac{3}{5} + 6\tfrac{2}{5} = 6\tfrac{2}{5} + 4\tfrac{3}{5}$$

   A. Associative

   B. Commutative

   C. Distributive

   D. Identity

2. Which property is used to rewrite the sum?

   $$2\tfrac{5}{8} + 0 = 2\tfrac{5}{8}$$

   F. Associative

   G. Commutative

   H. Distributive

   I. Identity

3. Which property is used to rewrite the sum?

   $$\left(3\tfrac{1}{4} + 2\tfrac{2}{3}\right) + 1\tfrac{1}{6} = 3\tfrac{1}{4} + \left(2\tfrac{2}{3} + 1\tfrac{1}{6}\right)$$

   A. Associative

   B. Commutative

   C. Distributive

   D. Identity

4. Which property is used to rewrite the sum?

   $$2\tfrac{1}{5} + 3\tfrac{7}{10} + 1\tfrac{1}{5} = 2\tfrac{1}{5} + 1\tfrac{1}{5} + 3\tfrac{7}{10}$$

   F. Associative

   G. Commutative

   H. Distributive

   I. Identity

PART 2 • Short Response

Record your answers in the space provided.

5. What is the denominator of the number that makes the number sentence below true?

   $$\tfrac{3}{4} + 2\tfrac{1}{2} = 2\tfrac{1}{2} + \square$$

   _____

6. What number makes the number sentence below true? Identify the property used.

   $$\left(\tfrac{2}{3} + \tfrac{3}{4}\right) + \tfrac{1}{2} = \tfrac{2}{3} + \left(\square + \tfrac{1}{2}\right)$$

   _____

PART 3 • Extended Response

 Record your answer in the space provided.

7. Find the solution to the problem below using the Associative and Commutative Properties. Justify your steps.

   $$\left(2\tfrac{3}{4} + 3\tfrac{1}{2}\right) + 5\tfrac{1}{4}$$

   _____

Lesson 12.1

PART 1 • Multiple Choice

Choose the best answer.

1.
$$\begin{array}{r} \frac{15}{16} \\ -\frac{7}{16} \\ \hline \end{array}$$

   A. $\frac{7}{16}$ C. $\frac{5}{8}$

   B. $\frac{3}{4}$ D. $\frac{1}{2}$

2.
$$\begin{array}{r} 7\frac{5}{12} \\ -3\frac{11}{12} \\ \hline \end{array}$$

   F. $4\frac{1}{2}$ H. $3\frac{1}{2}$

   G. $3\frac{2}{3}$ I. $3\frac{1}{3}$

3. $\frac{9}{10} - \frac{3}{10} =$

   A. $\frac{6}{1}$ C. $\frac{6}{20}$

   B. $\frac{3}{5}$ D. $\frac{1}{6}$

4. What value of a makes the number sentence below true?

$$a - \frac{5}{9} = \frac{3}{9}$$

   F. $\frac{8}{9}$ H. $\frac{2}{9}$

   G. $\frac{8}{18}$ I. $\frac{1}{9}$

5. What value of b makes the number sentence below true?

$$8\frac{11}{16} - b = 5\frac{1}{16}$$

   A. $13\frac{12}{16}$ C. $3\frac{10}{16}$

   B. $3\frac{12}{16}$ D. $3\frac{12}{32}$

PART 2 • Short Response

Record your answers in the space provided.

6. A board that was $12\frac{3}{8}$ inches long had a piece $9\frac{7}{8}$ inches long cut off. What is the whole number portion of the amount that is left?

_____

7. Show the steps needed to compare the two expressions and tell whether they are $>$, $<$, or $=$.

$$5\frac{2}{5} - \frac{3}{5} \bigcirc 4\frac{1}{5} + \frac{2}{5}$$

_____

_____

PART 3 • Extended Response

 Record your answer in the space provided.

8. Lori bought $8\frac{3}{8}$ pounds of potatoes and used $6\frac{7}{8}$ pounds in a casserole. She still needs 2 pounds to thicken her soup. Does Lori have enough potatoes left to make her soup? Explain.

_____

_____

Lesson 12.2

PART 1 • Multiple Choice

Choose the best answer.

1. $\frac{7}{8} - \frac{1}{2} =$

   A. $\frac{6}{6}$ C. $\frac{3}{8}$

   B. $\frac{0}{3}$ D. $\frac{2}{8}$

2. $\frac{1}{3} - \frac{1}{6} =$

   F. $\frac{0}{3}$ H. $\frac{1}{6}$

   G. $\frac{1}{3}$ I. $\frac{1}{9}$

3. $\frac{5}{6} - \frac{3}{4} =$

   A. $\frac{1}{24}$ C. $\frac{1}{8}$

   B. $\frac{1}{12}$ D. $\frac{1}{6}$

4. Caitlin lives $\frac{3}{4}$ mile from her school and $\frac{5}{8}$ mile from the library. How much farther is Caitlin's home from her school than from the library?

   F. $1\frac{3}{8}$ mile H. $\frac{1}{4}$ mile

   G. $\frac{1}{2}$ mile I. $\frac{1}{8}$ mile

5. Juan spent $\frac{1}{2}$ hour studying social studies and $\frac{1}{6}$ hour playing his harmonica. How much more time did Juan spend studying than playing?

   A. $\frac{1}{3}$ hour C. $\frac{1}{6}$ hour

   B. $\frac{2}{3}$ hour D. $\frac{1}{8}$ hour

PART 2 • Short Response

Record your answers in the space provided.

6. Ellie subtracted $\frac{3}{10}$ from $\frac{7}{8}$ and wrote her answer as a fraction in simplest form. What was Ellie's numerator?

   _____

7. Wei has a gold coin that is $\frac{3}{16}$ cm thick and a silver coin that is $\frac{5}{12}$ cm thick. How much thicker is the silver coin?

   _____

PART 3 • Extended Response

Record your answer in the space provided.

8. Sheets of plywood come in thicknesses of $\frac{1}{2}$ inch, $\frac{1}{4}$ inch, $\frac{3}{4}$ inch, $\frac{3}{8}$ inch, and $\frac{5}{8}$ inch. Which sheet is thickest? How much thicker is it than the next thickest? Show your steps.

   _____

   _____

   _____

Lesson 12.3

PART 1 • Multiple Choice

Choose the best answer.

1. Subtract.

    $7\frac{1}{2}$

    $-3\frac{1}{3}$

    A. $4\frac{1}{6}$ C. $4\frac{1}{4}$

    B. $4\frac{1}{5}$ D. $4\frac{1}{3}$

2. Subtract. $8\frac{11}{16} - 6\frac{5}{8} =$

    F. $2\frac{1}{16}$ H. $2\frac{3}{8}$

    G. $2\frac{3}{16}$ I. $2\frac{3}{4}$

3. $3\frac{3}{4} - \frac{1}{2} =$

    A. $3\frac{2}{2}$ C. $3\frac{1}{4}$

    B. $3\frac{1}{2}$ D. $3\frac{1}{6}$

4. A recipe requires 4 cups of flour. Janice has $3\frac{3}{8}$ cups of flour in a bowl. How much more flour must she add to the bowl for this recipe?

    F. $1\frac{7}{8}$ cup H. $\frac{7}{8}$ cup

    G. $1\frac{5}{8}$ cup I. $\frac{5}{8}$ cup

5. Tim walked from Gilbert Lake to his cabin. The cabin is $3\frac{1}{4}$ miles from the lake. After $1\frac{2}{3}$ miles, Tim stops to have a snack. How much farther must he walk at this point?

    A. $4\frac{11}{12}$ miles C. $1\frac{2}{7}$ miles

    B. $2\frac{7}{12}$ miles D. $1\frac{7}{12}$ miles

PART 2 • Short Response

Record your answers in the space provided.

6. The price of a share of stock rose from $\$8\frac{1}{2}$ to $\$12\frac{1}{4}$. By how much did the price rise? Write your answer as a decimal.

    _____

7. A can of tuna fish contains $6\frac{1}{2}$ ounces of tuna. Harry used $2\frac{3}{4}$ ounces for his sandwich. How much tuna fish is left? Write your answer as a mixed number in simplest form.

    _____

PART 3 • Extended Response

 Record your answer in the space provided.

8. Describe the steps you would take to find $5\frac{7}{10} - 1\frac{1}{4}$.

    _____

    _____

    _____

Lesson 12.4

PART 1 • Multiple Choice

Choose the best answer.

1. $5\frac{1}{2} - 2\frac{5}{6}$

   A. $2\frac{1}{3}$ C. $3\frac{1}{3}$

   B. $2\frac{2}{3}$ D. $3\frac{2}{3}$

2. $10\frac{1}{3} - 4\frac{7}{10} =$

   F. $6\frac{19}{30}$ H. $5\frac{19}{30}$

   G. $6\frac{11}{30}$ I. $5\frac{11}{30}$

3. What value of g makes the number sentence below true?

   $8\frac{1}{4} - g = 7\frac{1}{8}$

   A. $1\frac{1}{8}$ C. $15\frac{1}{8}$

   B. $1\frac{3}{8}$ D. $15\frac{3}{8}$

4. Kendra used $3\frac{1}{2}$ pounds of Granny Smith apples and $1\frac{3}{4}$ pounds of Fuji apples to bake pies. How many more pounds of Granny Smith apples than Fuji apples did Kendra use?

   F. $1\frac{1}{4}$ H. $2\frac{1}{4}$

   G. $1\frac{3}{4}$ I. $2\frac{3}{4}$

5. A 100-acre property was divided into 16 equal lots. Each lot measured $6\frac{1}{4}$ acres. Paula bought a lot and cleared $2\frac{7}{8}$ acres. How much of her lot was left uncleared?

   A. $4\frac{4}{5}$ acres   C. $3\frac{5}{8}$ acres

   B. $4\frac{3}{8}$ acres   D. $3\frac{3}{8}$ acres

PART 2 • Short Response

Record your answers in the space provided.

6. A piece of cloth measuring $3\frac{1}{2}$ yards long was cut from a bolt of cloth measuring 18 yards long. How many yards long is the cloth that remained on the bolt? Write your answer as a decimal.

   _____

7. A refrigerator is $28\frac{7}{8}$ inches wide. The opening in Javier's kitchen where the refrigerator must go is $29\frac{3}{16}$ inches wide. How much wider is the opening than the refrigerator?

   _____

PART 3 • Extended Response

 Record your answer in the space provided.

8. How is subtracting mixed numbers like subtracting fractions? How is it different?

   _____

   _____

Lesson 12.5

PART 1 • Multiple Choice

Choose the best answer.

1. Latisha's puppy weighed $3\frac{1}{4}$ pounds when she brought it home. Now the puppy weighs $8\frac{7}{8}$ pounds. How much weight did the puppy gain?

   A. $12\frac{1}{4}$ pounds **C.** $5\frac{5}{8}$ pounds

   B. $11\frac{2}{3}$ pounds **D.** $5\frac{3}{4}$ pounds

2. George was $47\frac{3}{4}$ inches tall. During the year, he grew $2\frac{1}{2}$ inches. What was his height at the end of the year?

   F. $45\frac{1}{4}$ inches **H.** $49\frac{2}{3}$ inches

   G. $49\frac{1}{4}$ inches **I.** $50\frac{1}{4}$ inches

3. The Pine Trail is $3\frac{5}{16}$ miles long. After walking $1\frac{2}{3}$ miles of the Pine Trail, Mel turned around and headed back to camp. How far did Mel walk in all?

   A. $1\frac{2}{3}$ miles **C.** $2\frac{1}{3}$ miles

   B. $1\frac{31}{48}$ miles **D.** $3\frac{1}{3}$ miles

4. Zena bought 1 pound of mixed nuts. She had $\frac{1}{4}$ pound of cashews and $\frac{1}{3}$ pound of walnuts. The rest was almonds. What amount of almonds did Zena buy?

   F. $1\frac{7}{12}$ pounds **H.** $\frac{5}{12}$ pound

   G. $\frac{7}{12}$ pounds **I.** $\frac{5}{7}$ pound

5. Stu worked 55 hours. $37\frac{1}{2}$ hours were regular time. The rest was overtime. How many hours of overtime did Stu work?

   A. $17\frac{1}{2}$ hours **C.** $27\frac{1}{2}$ hours

   B. $18\frac{1}{2}$ hours **D.** $92\frac{1}{2}$ hours

PART 2 • Short Response

Record your answers in the space provided.

6. At the shallow end of a pool, the depth is $3\frac{3}{4}$ feet. At the deep end of the pool, the depth is 10 feet. How many feet deeper is the deep end? Write your answer as a decimal.

   _____

7. A prescription medicine comes in two dosages: $\frac{5}{8}$ mg and $1\frac{1}{4}$ mg. How much greater is the larger dosage? Write your answer as a fraction or mixed number in simplest form.

   _____

PART 3 • Extended Response

Record your answer in the space provided.

8. Make up a problem of your own that can be solved using the strategy of writing an equation. Show the equation you would use to solve the problem.

   _____

   _____

   _____

Lesson 12.6

PART 1 • Multiple Choice

Choose the best answer.

1. When rounded to the nearest whole number, which mixed number rounds to 10?

   A. $10\frac{4}{5}$ C. $9\frac{5}{9}$

   B. $10\frac{2}{3}$ D. $9\frac{1}{3}$

2. When rounded to the nearest whole number, which mixed number rounds to 6?

   F. $6\frac{9}{20}$ H. $5\frac{2}{5}$

   G. $6\frac{3}{4}$ I. $5\frac{3}{10}$

3. Which of the following is the best estimate of $3\frac{2}{3} + 7\frac{1}{5}$?

   A. 10   C. 12

   B. 11   D. 13

4. Which of the following is the best estimate of $43\frac{7}{8} - 28\frac{1}{40}$?

   F. 26   H. 16

   G. 25   I. 15

5. Which of the following is the best estimate of $9\frac{5}{6} + 8\frac{5}{6}$?

   A. 17   C. 19

   B. 18   D. 20

PART 2 • Short Response

Record your answers in the space provided.

6. Janice estimated $148\frac{1}{2} - 52\frac{3}{4}$ by rounding each number to the nearest ten. What was Janice's estimate?

   _____

7. Carl needs pieces of framing that measure $8\frac{5}{8}$ inches, $12\frac{3}{4}$ inches and $3\frac{7}{8}$ inches. To estimate the total amount he needs, Carl rounded each measure to the nearest whole number and added. What was Carl's estimate of the total?

   _____

PART 3 • Extended Response

 Record your answer in the space provided.

8. Show how rounding mixed numbers to the nearest whole number or the nearest ten can lead to different estimates when finding a sum. Create your own example to show this.

   _____

   _____

Lesson 13.1

PART 1 • Multiple Choice

Choose the best answer.

1. $\frac{2}{3} \times 24 =$

   A. 4   C. 12

   B. 8   D. 16

2. $50 \times \frac{7}{10} =$

   F. 350   H. 35

   G. 70   I. 12

3. $\frac{3}{8} \times 72 =$

   A. 192   C. 24

   B. 27   D. 12

4. Carmen has read $\frac{3}{4}$ of her 200 page book. How many pages has she read?

   F. 150   H. 75

   G. 125   I. 53

5. In a display of 126 circles, $\frac{2}{3}$ of them contain the color red. How many circles contain the color red?

   A. 42   C. 84

   B. 63   D. 115

PART 2 • Short Response

Use the table for problems 6–7.

Rule: Multiply by $\frac{8}{5}$	
Input	Output
35	56
50	80
90	

Record your answers in the space provided.

6. What is the missing value for the table?

   _____

7. If you wanted a whole number output, should you use 25 or 48 as the input? Explain.

   _____

PART 3 • Extended Response

 Record your answer in the space provided.

8. What is the rule for the table below? Explain.

Rule: Multiply by ☐	
Input	Output
21	3
42	6
140	20

   _____

Lesson 13.2

PART 1 • Multiple Choice

Choose the best answer.
Use the model shown below for problems 1-5.

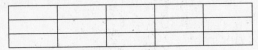

1. $\frac{1}{5}$ of $\frac{1}{3}$ is _____

   A. $\frac{1}{8}$ C. $\frac{1}{9}$

   B. $\frac{2}{8}$ D. $\frac{1}{15}$

2. $\frac{2}{3}$ of $\frac{1}{5}$ is _____

   F. $\frac{1}{15}$ H. $\frac{3}{15}$

   G. $\frac{2}{15}$ I. $\frac{4}{15}$

3. $\frac{3}{5}$ of $\frac{1}{3}$ is _____

   A. $\frac{1}{15}$ C. $\frac{3}{15}$

   B. $\frac{2}{15}$ D. $\frac{4}{15}$

4. $\frac{2}{5}$ of $\frac{2}{3}$ is _____

   F. $\frac{8}{15}$ H. $\frac{4}{15}$

   G. $\frac{6}{15}$ I. $\frac{2}{15}$

5. $\frac{2}{3}$ of $\frac{4}{5}$ is _____

   A. $\frac{12}{15}$ C. $\frac{4}{15}$

   B. $\frac{8}{15}$ D. $\frac{2}{15}$

PART 2 • Short Response

Record your answers in the space provided. Use the model shown below for problems 6 and 7.

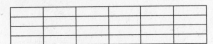

6. Find $\frac{5}{6}$ of $\frac{3}{4}$. Write your answer as a decimal rounded to the nearest hundredth.

   _____

7. Find $\frac{1}{4}$ of $\frac{1}{6}$. Write your answer as a fraction.

   _____

PART 3 • Extended Response

Record your answer in the space provided.

8. Find $\frac{2}{3}$ of $\frac{3}{5}$ using a model that you draw and complete.

   _____

   _____

Lesson 13.3

PART 1 • Multiple Choice

Choose the best answer.

1. How is the product $\frac{2}{5} \times \frac{1}{4}$ written in simplest form?

   A. $\frac{3}{20}$ C. $\frac{1}{10}$

   B. $\frac{8}{20}$ D. $\frac{1}{3}$

2. How is the product $\frac{3}{8} \times \frac{5}{6}$ written in simplest form?

   F. $\frac{1}{6}$ H. $\frac{4}{7}$

   G. $\frac{15}{14}$ I. $\frac{5}{16}$

3. How is the product $\frac{5}{12} \times \frac{3}{5}$ written in simplest form?

   A. $\frac{1}{4}$ C. $\frac{15}{17}$

   B. $\frac{8}{17}$ D. $\frac{15}{60}$

4. One-half of the students in a school are girls. One-seventh of the girls are in the 5th grade. What fraction of the students in the school are 5th-grade girls?

   F. $\frac{1}{9}$ H. $\frac{1}{14}$

   G. $\frac{2}{9}$ I. $\frac{2}{14}$

5. Ed has a garden. He grows vegetables in two-thirds of his garden. One-eighth of Ed's vegetable garden is used for growing tomatoes. What fraction of Ed's home garden is used for growing tomatoes?

   A. $\frac{2}{11}$ C. $\frac{1}{24}$

   B. $\frac{1}{12}$ D. $\frac{19}{24}$

PART 2 • Short Response

Record your answers in the space provided.

6. How is the product $\frac{7}{18} \times \frac{9}{14}$ written as a decimal?

   _____

7. Arielle worked on $\frac{5}{6}$ of the days in June. She spent $\frac{3}{5}$ of those days working in the company warehouse. On what fraction of the days in June did Arielle work in the company warehouse? Write your answer as a fraction in simplest form.

   _____

PART 3 • Extended Response

Record your answer in the space provided.

8. Tony folded a piece of paper in half, cut it along the fold, and kept one of the two pieces. He then folded his new piece of paper, cut it along the fold, and kept one of the two pieces. In all, he did four folds and cuts, and kept one of the last pieces. How big is Tony's last piece, compared to the original piece of paper? Explain.

   _____

   _____

Lesson 13.4

PART 1 • Multiple Choice

Choose the best answer.

1. A rectangular poster has a height of $\frac{3}{4}$ m and a width of $\frac{1}{2}$ m. How much framing material would be needed for this poster?

   A. $\frac{3}{8}$ m   C. $1\frac{1}{2}$ m

   B. $1\frac{1}{4}$ m   D. $2\frac{1}{2}$ m

2. Twice this week Amanda spent $\frac{2}{3}$ of her work day sewing a robe and $\frac{1}{3}$ of one work day sewing a skirt. How much time in all did Amanda spend sewing the robe and skirt this week?

   F. 2 work days

   G. $1\frac{2}{3}$ work days

   H. $1\frac{1}{3}$ work days

   I. 1 work day

3. A trail has a hilly section that is $\frac{5}{8}$ of a mile long and a flat section that is $\frac{1}{4}$ of a mile long. If Scott hikes the trail once in each direction, how far does he hike in all?

   A. $\frac{7}{8}$ mile   C. $1\frac{3}{4}$ mile

   B. $1\frac{1}{3}$ mile   D. $1\frac{7}{8}$ mile

4. Rene used $1\frac{1}{2}$ cups of flour to make a tart and $\frac{3}{4}$ cup of flour to make eight pancakes. How much flour did Rene use in all?

   F. $1\frac{2}{3}$ cups   H. 3 cups

   G. $2\frac{1}{4}$ cups   I. $7\frac{1}{2}$ cups

PART 2 • Short Response

Record your answers in the space provided.

5. A picture is 12 inches high. Its frame is $\frac{3}{4}$ of an inch wide at all points. How many inches high are the picture and frame together? Write your answer as a decimal.

   _____

6. Lin spent 6 hours at school Wednesday. During that time, he spent $\frac{2}{3}$ of an hour at lunch and $\frac{2}{3}$ of an hour at an assembly. The rest of the time, Lin was in class. How much time did Lin spend in class? Write your answer as a mixed number.

   _____

PART 3 • Extended Response

Record your answer in the space provided.

7. A DVD player is 17 inches wide. For shipping, the player is protected on each side by foam that is $\frac{3}{4}$ of an inch thick. It is then placed in a carton that is $\frac{1}{8}$ of an inch thick on each side. How wide is the carton? Show your method.

   _____

   _____

   _____

Lesson 13.5

PART 1 • Multiple Choice

Choose the best answer.

1. Which of the following is the best estimate of $\frac{1}{3} \times 11$?

   A. 6   C. 3

   B. 4   D. 2

2. Which of the following is the best estimate of $4\frac{4}{5} \times 5\frac{1}{4}$?

   F. 15   H. 25

   G. 20   I. 30

3. Which of the following is the best estimate of $31\frac{1}{8} \times \frac{2}{3}$?

   A. 20   C. 10

   B. 15   D. 5

4. Ira works $6\frac{3}{4}$ hours per day and 5 days per week. Which of the following is the best estimate of the number of hours Ira works each week?

   F. 20   H. 30

   G. 25   I. 35

5. There are 716 students in a school. Three-fourths of the students bring their lunch to school. Which of the following is the best estimate of the number of students that bring their lunch to school?

   A. 180   C. 540

   B. 500   D. 600

PART 2 • Short Response

Record your answers in the space provided.

6. To estimate the product $\frac{5}{8} \times 46$, Tina used compatible numbers. With which number did Tina replace 46?

   _____

7. To estimate the product $123\frac{1}{2} \times 67\frac{7}{8}$, Jerry rounded each factor to the nearest 10 and then multiplied. What was Jerry's estimate?

   _____

PART 3 • Extended Response

Record your answer in the space provided.

8. A newspaper article reported that $\frac{1}{20}$ of the residents in a town of 39,850 were over age 80. Justin estimated this number as follows: Since $\frac{1}{20}$ is very close to 0, $\frac{1}{20}$ of 39,850 is approximately $0 \times 39,850$, which equals 0. Did Justin estimate well? Explain your thinking.

   _____

   _____

Lesson 13.6

PART 1 • Multiple Choice

Choose the best answer.

1. $\frac{2}{3} \times \frac{3}{4} =$

   A. $\frac{5}{7}$ C. $\frac{5}{12}$

   B. $\frac{1}{2}$ D. $\frac{1}{6}$

2. $\frac{4}{5} \times 60 =$

   F. 75   H. 44

   G. 48   I. 16

3. $\frac{4}{7} \times \frac{5}{9} =$

   A. $\frac{20}{16}$ C. $\frac{9}{16}$

   B. $\frac{35}{36}$ D. $\frac{20}{63}$

4. $\frac{2}{3} \times \frac{1}{4} \times 24 =$

   F. 24   H. 4

   G. 6   I. $\frac{1}{4}$

5. What value of n makes the number sentence below true?

   $n \times \frac{5}{9} = 100$

   A. 180   C. 500

   B. 360   D. 900

PART 2 • Short Response

Record your answers in the space provided.

6. You sold $\frac{7}{10}$ of your coin collection. If you had 340 coins, how many did you sell?

   _____

7. Find $\frac{2}{5}$ of 30 and draw a model.

   _____

PART 3 • Extended Response

Record your answer in the space provided.

8. If you know that $\frac{1}{8}$ of a number is 30, what is $\frac{7}{8}$ of the number? Explain.

   _____

   _____

   _____

Lesson 14.1

PART 1 • Multiple Choice

Choose the best answer.

1. How is the product $3\frac{1}{2} \times 5$ written in simplest form?

   A. $15\frac{1}{10}$ C. $16\frac{1}{2}$

   B. $15\frac{1}{2}$ D. $17\frac{1}{2}$

2. How is the product $6 \times 2\frac{1}{5}$ written in simplest form?

   F. $13\frac{2}{5}$ H. $12\frac{1}{30}$

   G. $13\frac{1}{5}$ I. $2\frac{1}{30}$

3. How is the product $32\frac{2}{3} \times 6$ written in simplest form?

   A. 186   C. 196

   B. $194\frac{2}{3}$ D. 201

4. Janice worked $2\frac{1}{2}$ hours each day for 4 days. If she earns $24 per hour, how much did she earn over the four days?

   F. $120   H. $200

   G. $160   I. $240

5. What value of w makes the number sentence below true?

   $w \times 4\frac{1}{3} = 30\frac{1}{3}$

   A. 7   C. 12

   B. 8   D. 26

PART 2 • Short Response

Record your answers in the space provided.

6. Bruce ordered 140 poles and $4\frac{1}{2}$ times as many rails to build a fence. How many rails did he order?

   _____

7. Show how to multiply $3\frac{1}{2}$ by 6 using the Distributive Property.

   _____

PART 3 • Extended Response

Record your answer in the space provided.

8. Explain how to make a model to show the multiplication of $5 \times 2\frac{3}{5}$ and give the answer in simplest form.

   _____

   _____

   _____

   _____

Lesson 14.2

PART 1 • Multiple Choice

Choose the best answer.

1. How is the product $\frac{1}{2} \times 3\frac{1}{2}$ written in simplest form?

   A. $3\frac{1}{4}$ C. $1\frac{3}{4}$

   B. 2 D. $1\frac{1}{2}$

2. How is the product $\frac{5}{7} \times 2\frac{3}{5}$ written in simplest form?

   F. $2\frac{15}{35}$ H. $1\frac{6}{7}$

   G. $2\frac{8}{35}$ I. $1\frac{3}{7}$

3. How is the product $\frac{7}{10} \times 2\frac{2}{9}$ written in simplest form?

   A. $1\frac{4}{9}$ C. $1\frac{56}{90}$

   B. $1\frac{5}{9}$ D. $2\frac{7}{45}$

4. Maria has a piece of fabric $6\frac{2}{3}$ yards long. She will use $\frac{3}{4}$ of it for a gown. How much fabric will she use for the gown?

   F. $4\frac{3}{4}$ yd   H. $5\frac{1}{6}$

   G. 5 yd   I. $5\frac{1}{4}$ yd

5. A program about World War II is $18\frac{1}{2}$ hours long. Buck watched $\frac{2}{3}$ of the program. How many hours of the program did Buck watch?

   A. $12\frac{1}{3}$ hr   C. $12\frac{2}{3}$ hr

   B. $12\frac{1}{2}$ hr   D. $12\frac{5}{6}$ hr

PART 2 • Short Response

Record your answers in the space provided.

6. How is the product $10\frac{4}{5} \times \frac{5}{8}$ written as a decimal?

   _____

7. A bookshelf is $12\frac{3}{4}$ inches wide. Mae's books fill $\frac{1}{3}$ of the shelf. How many inches of the shelf do Mae's books occupy? Write your answer as a mixed number in simplest form.

   _____

PART 3 • Extended Response

Record your answer in the space provided.

8. Show how you could use the Distributive Property to find $\frac{4}{5} \times 8\frac{1}{3}$.

   _____

   _____

   _____

   _____

Lesson 14.3

PART 1 • Multiple Choice

Choose the best answer.

1. How is the product $1\frac{1}{4} \times 1\frac{1}{4}$ written in simplest form?

    A. $1\frac{1}{16}$ C. $2\frac{1}{16}$

    B. $1\frac{9}{16}$ D. $2\frac{1}{2}$

2. How is the product $8\frac{1}{3} \times 4\frac{4}{5}$ written in simplest form?

    F. $32\frac{4}{15}$ H. 40

    G. $39\frac{2}{5}$ I. 50

3. How is the product $4\frac{1}{6} \times 1\frac{4}{5}$ written in simplest form?

    A. $4\frac{2}{15}$ C. $7\frac{1}{2}$

    B. $6\frac{29}{30}$ D. $9\frac{3}{8}$

4. How is the product $7\frac{1}{3} \times 2\frac{5}{11}$ written in simplest form?

    F. $14\frac{5}{33}$ H. $17\frac{10}{33}$

    G. $16\frac{4}{11}$ I. 18

5. Ari is multiplying a recipe by $1\frac{1}{2}$. The recipe normally uses $3\frac{1}{4}$ cups of flour. How many cups of flour will Ari use?

    A. $4\frac{7}{8}$ C. $3\frac{3}{4}$

    B. $4\frac{3}{4}$ D. $3\frac{1}{8}$

PART 2 • Short Response

Record your answers in the space provided.

6. Kristin rode her bike for $2\frac{1}{2}$ hours. Her average speed was $12\frac{1}{2}$ miles per hour. How many miles did Kristin ride? Write your answer as a decimal.

    _____

7. A jogging path around a reservoir is $1\frac{5}{8}$ miles long. Justin ran $3\frac{1}{2}$ times around the reservoir. How many miles did Justin run? Write your answer as a mixed number in simplest form.

    _____

PART 3 • Extended Response

Record your answer in the space provided.

8. Explain how you can use mental math to tell that $9\frac{1}{4} \times 5\frac{7}{9}$ must be greater than 45 and less than 60.

    _____

    _____

Lesson 14.4

PART 1 • Multiple Choice

Choose the best answer.
Franco finds the product of the numbers on which each spinner below lands. Use this information for problems 1 and 2. Make an organized list to solve the problems.

1. What is the least product possible?

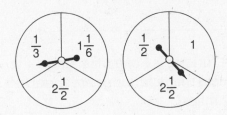

   A. $\frac{1}{6}$ C. $\frac{1}{2}$
   B. $\frac{1}{3}$ D. $\frac{5}{6}$

2. How many different products are there?

   F. 6   H. 8
   G. 7   I. 9

Kiera throws two darts. For each dart, she gets the points indicated. Her total score is the sum of the scores for each of her two darts. Use this information for problems 3 and 4.

3. How many different total scores are there?

   A. 3   C. 8
   B. 6   D. 9

4. How many possible totals are even?

   F. 1   H. 3
   G. 2   I. 4

PART 2 • Short Response

Record your answers in the space provided. In the first round of a game, each player selects a red, white or blue chip. In the second round of the game, each player selects a red, orange, green, or yellow chip. Use this information for problems 5 and 6.

5. How many different combinations of chips can be chosen by a player?

   _____

6. How many different combinations of chips in which neither chip is red can be chosen by a player?

   _____

PART 3 • Extended Response

THINK
SOLVE
EXPLAIN

Record your answer in the space provided.

7. The tens digit of a two-digit number is odd. The ones digit is 5, 6, or 7. Show how you can make an organized list to find all the numbers that fit these two requirements.

   _____

   _____

Lesson 14.5

PART 1 • Multiple Choice

Choose the best answer.

1. $3 \div \frac{1}{6} =$

   A. $\frac{1}{2}$ C. 9

   B. $2\frac{5}{6}$ D. 18

2. $13 \div \frac{1}{5} =$

   F. 65   H. 8

   G. 55   I. $2\frac{3}{5}$

3. $8 \div \frac{1}{8} =$

   A. 72   C. 16

   B. 64   D. 1

4. Each of Cal's chocolates weighs $\frac{1}{3}$ ounce. How many chocolates are there in a 6-ounce box?

   F. 6   H. 18

   G. 12   I. 24

5. Maria uses $\frac{1}{2}$ teaspoon of vanilla when she makes hot cocoa. There are 12 teaspoons left in her bottle of vanilla. How many times can Maria make hot cocoa?

   A. 12   C. 24

   B. 13   D. 36

PART 2 • Short Response

Record your answers in the space provided.

6. A track is $\frac{1}{4}$ mile long. Wendell wants to run 4 miles on this track. How many times must he run around the track?

   _____

7. A furlong is $\frac{1}{5}$ of a mile long. How many furlongs are there in 2 miles?

   _____

PART 3 • Extended Response

 Record your answer in the space provided.

8. Roc made three trays of lasagna for a party. He estimates that a person eats $\frac{1}{8}$ of a tray. How many people can Roc feed with 3 trays of lasagna? Draw a diagram that helps explain your reasoning.

   _____

   _____

   _____

Lesson 14.6

PART 1 • Multiple Choice

Choose the best answer.

1. What is the reciprocal of $\frac{5}{12}$?

   A. $\frac{12}{5}$ C. $\frac{5}{12}$

   B. $-\frac{12}{5}$ D. $-\frac{5}{12}$

2. $\frac{3}{8} \div \frac{1}{4} =$

   F. $\frac{3}{32}$ H. $\frac{1}{3}$

   G. $\frac{3}{12}$ I. $\frac{3}{2}$

3. $15 \div \frac{3}{5} =$

   A. 9   C. $15\frac{3}{5}$

   B. 10   D. 25

4. What value of n makes the number sentence true?

   $\frac{3}{5} \div n = \frac{3}{10}$

   F. 5   H. $\frac{1}{2}$

   G. 2   I. $\frac{3}{10}$

5. Anna's piano lessons are $\frac{3}{4}$ of an hour long. Last year, Anna spent 36 hours in piano lessons. How many lessons did Anna take last year?

   A. 24   C. 48

   B. 42   D. 54

PART 2 • Short Response

Record your answers in the space provided.

6. Renaldo is making seat covers. Each seat cover requires $\frac{3}{8}$ of a yard of fabric. If Renaldo has 12 yards of fabric, how many seat covers can he make?

   _____

7. $\frac{1}{4} \div \frac{2}{7} =$

   _____

PART 3 • Extended Response

 Record your answer in the space provided.

8. A teaspoon equals $\frac{1}{6}$ of a fluid ounce. A cup equals 8 fluid ounces. How many teaspoons are there in 1 cup? Show your steps.

   _____

   _____

   _____

Lesson 15.1

PART 1 • Multiple Choice

Choose the best answer.

1. Complete the sentence:
   120 minutes = _____.

   A. 2 seconds   C. 2,880 seconds

   B. 720 seconds   D. 7,200 seconds

2. Complete the sentence:
   50 months = _____.

   F. 600 years

   G. 12 years 2 months

   H. 5 years

   I. 4 years 2 months

3. What is the elapsed time from
   2:15 A.M. to 8:00 A.M.?

   A. 5 hours 45 minutes

   B. 6 hours 45 minutes

   C. 6 hours 45 minutes

   D. 10 hours 15 minutes

4. What is the elapsed time from
   7:02 P.M. to 11:23 A.M.?

   F. 4 hours 21 minutes

   G. 14 hours 21 minutes

   H. 16 hours 21 minutes

   I. 17 hours 21 minutes

5. Choose the most reasonable unit of
   time to complete the following
   sentence: It took Leah 10 ▨ to walk
   to school this morning.

   A. days   C. seconds

   B. minutes   D. years

PART 2 • Short Response

Record your answers in the space
provided.

6. How many hours are there
   in 31 days?

   _____

7. Nelson started work today
   at 8:45 A.M. and finished at
   4:30 P.M. How much time
   did Nelson spend at work?

   _____

PART 3 • Extended Response

Record your answer in the space
provided.

8. From the time she was 4 years old, In
   Sook has practiced the violin every
   day for 60 minutes. In Sook is now 10
   years old. Estimate how much time
   she has spent practicing the violin in
   her life. Explain your method. Express
   your answer in months.

   _____

   _____

   _____

Lesson 15.2

PART 1 • Multiple Choice

Choose the best answer.
Emily used steps to measure various lengths. Her classroom was 20 steps long. Her porch was 10 steps long. Her living room was 15 steps long. The auditorium was 90 steps long. Her kitchen was 8 steps long. Use this information for problems 1–3.

1. Which of the following had the greatest length?

   A. auditorium   C. living room

   B. kitchen   D. porch

2. Which of the following had the shortest length?

   F. auditorium   H. kitchen

   G. classroom   I. living room

3. Complete the statement: Emily's classroom is about _____ as Emily's porch.

   A. five times as long

   B. half as long

   C. ten times as long

   D. twice as long

4. Which of the following is most reasonably measured using hand widths?

   F. the height of a building

   G. the length of a table

   H. the thickness of a sheet of construction paper

   I. the weight of a basketball

PART 2 • Short Response

Record your answers in the space provided.

5. Chuck measured the width of a bookcase using large paper clips. The width was 44 paper clips. If each paper clip is $1\frac{1}{2}$ inches long, how many inches wide is the bookcase?

   _____

6. Kaitlin used her textbook to measure the width of her piano. The textbook is 10 inches long, and the piano was 6 textbooks wide. What is the width of the piano, in inches?

   _____

PART 3 • Extended Response

Record your answer in the space provided.

7. Describe a way you could use nonstandard measures to compare the widths of two computer screens.

   _____

   _____

   _____

Lesson 15.3

PART 1 • Multiple Choice

Choose the best answer.
Use the pencil below for problems 1 and 2.

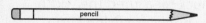

1. What is the length of the pencil, to the nearest inch?

   A. 1 inch   **C.** $1\frac{3}{4}$ inches

   B. $1\frac{1}{2}$ inches   **D.** 2 inches

2. What is the length of the pencil, to the nearest eighth inch?

   F. $2\frac{1}{8}$ inches   **H.** $1\frac{7}{8}$ inches

   G. 2 inches   **I.** $1\frac{3}{4}$ inches

3. Which unit is most appropriate to use to measure the distance between the capital cities of two states?

   A. foot   **C.** mile

   B. inch   **D.** yard

4. Complete the sentence:
   30 feet = _____.

   F. 90 yards   **H.** 10 yards

   G. 15 yards   **I.** 2.25 yards

5. Which is the greatest length?

   A. 1 mile   **C.** 3,500 feet

   B. 1,800 yards   **D.** 50,000 inches

PART 2 • Short Response

Record your answers in the space provided.

6. Victoria stands 4 feet 4 inches tall. How many inches tall is Victoria?

   _____

7. Each story of a building is 132 inches high. If the building is 33 stories high, what is its height, in yards?

   _____

PART 3 • Extended Response

 Record your answer in the space provided.

8. A running track is usually $\frac{1}{4}$ mile long. How many feet is that? How many yards is that? Show your method.

   _____

   _____

   _____

Lesson 15.4

PART 1 • Multiple Choice

Choose the best answer.

1. Which unit is most appropriate to measure the weight of a watermelon?

   A. cup   **C.** pound

   B. ounce   **D.** ton

2. Complete the sentence: 16,000 pounds = _____.

   F. 1,000 tons   **H.** 8 tons

   G. 80 tons   **I.** 2 tons

3. Complete the sentence: 100 fluid ounces = _____.

   A. 3 cups 4 fluid ounces

   B. 3 pints 4 fluid ounces

   C. 3 quarts 4 fluid ounces

   D. 3 gallons 4 fluid ounces

4. A typical hamburger weighs 4 ounces. How many hamburgers can be made from 3 pounds of hamburger meat?

   F. 24   **H.** 9

   G. 12   **I.** 6

5. For each cup of rice that Jim cooks, he uses 2 cups of water. Today, he is making 6 cups of rice for a group of friends. How much water must Jim use?

   A. 3 quarts   **C.** 3 gallons

   B. 3 pints   **D.** 6 quarts

PART 2 • Short Response

Record your answers in the space provided.

6. Fruit juice is sold in 1-gallon jugs and individual-sized containers that measure 4 fluid ounces. How many individual-sized containers must you buy to equal the amount in a 1-gallon jug?

   _____

7. A freight train carries 6 tons of coal. How many pounds of coal is this?

   _____

PART 3 • Extended Response

 Record your answer in the space provided.

8. A pint of water weighs 1 pound. How much does a 10-gallon jug of water weigh? Show your method.

   _____

   _____

   _____

Lesson 15.5

PART 1 • Multiple Choice

Choose the best answer.

1. Which is the most reasonable estimate for the length of a sports car?

   A. 14 feet   C. 60 feet

   B. 14 yards   D. 60 yards

2. Which is the most reasonable estimate for the weight of a shoe?

   F. 8 pounds   H. 48 pounds

   G. 8 ounces   I. 48 ounces

3. Which is the most reasonable estimate for the capacity of a sink?

   A. 3 fluid ounces

   B. 3 cups

   C. 3 pints

   D. 3 gallons

4. Which is the most reasonable estimate for the height of a one-story house?

   F. 25 feet   H. 25 miles

   G. 25 inches I. 25 yards

5. Which is the most reasonable estimate for the weight of a full-grown cat?

   A. 40 pounds C. 10 pounds

   B. 40 ounces D. 10 ounces

PART 2 • Short Response

Record your answers in the space provided.

6. Karen lives 2 miles from school. She says that the distance is about 10,000 feet. What is the actual distance, in yards?

   _____

7. The crankcase of Harry's car takes 9 pints of oil. He bought 5 quarts of oil at the auto store to fill his crankcase. How much oil will Harry have left over?

   _____

PART 3 • Extended Response

Record your answer in the space provided.

8. A group of 12 friends are skiing for 5 days. Vera is buying orange juice for the group to last the entire 5 days. She estimates that each skier will drink 8 fluid ounces per day. If orange juice is sold in half-gallons and gallons, what is a reasonable amount of orange juice for Vera to buy? Explain your thinking.

   _____

   _____

Lesson 16.1

PART 1 • Multiple Choice

Choose the best answer.
Use the segment shown below for problems 1–2.

_____

1. What is the length of the segment, to the nearest centimeter?

   A. 5 cm   **C.** 6 cm

   B. 5.8 cm   **D.** 58 cm

2. What is the length of the segment, to the nearest millimeter?

   F. 5 mm   **H.** 6 mm

   G. 5.8 mm   **I.** 58 mm

3. Which is the most appropriate unit for measuring the distance across a playground?

   A. centimeter   **C.** millimeter

   B. kilometer   **D.** meter

4. Which is the most appropriate unit for measuring the width of a quarter?

   F. centimeter   **H.** millimeter

   G. kilometer   **I.** meter

5. Which is the most appropriate unit for measuring your height?

   A. centimeter   **C.** millimeter

   B. kilometer   **D.** meter

PART 2 • Short Response

Record your answers in the space provided. Use the segment shown below for problems 6–7.

_____

6. What is the length of the segment, in millimeters?

_____

7. What is the length of the segment, to the nearest centimeter?

_____

PART 3 • Extended Response

 Record your answer in the space provided.

8. Would you measure the length of a truck in meters or kilometers? Explain your answer.

_____

_____

_____

Lesson 16.2

PART 1 • Multiple Choice

Choose the best answer.

1. Which is the most appropriate unit to measure the mass of a pencil?

   A. gram   C. liter

   B. kilogram   D. milliliter

2. Which is the most appropriate unit to measure the capacity of a bathtub?

   F. gram   H. liter

   G. kilogram   I. milliliter

3. Which measurement best fits a small dog?

   A. 50 kilograms

   B. 50 grams

   C. 5 kilograms

   D. 5 grams

4. Which measurement best fits a flower vase?

   F. 20 milliliters   H. 20 liters

   G. 1 liter   I. 100 liters

5. Which measurement best fits a cracker?

   A. 50 kilograms

   B. 50 grams

   C. 5 kilograms

   D. 5 grams

PART 2 • Short Response

Record your answers in the space provided.

6. How many centiliters are in 1 liter?

   _____

7. How many grams are in 1 kilogram?

   _____

PART 3 • Extended Response

Record your answer in the space provided.

8. Would you measure the water capacity of a washing machine in milliliters, centiliters, or liters? What unit of measure would you use to weigh the clothes going into the washing machine? Explain your answer.

   _____

   _____

   _____

Lesson 16.3

PART 1 • Multiple Choice

Choose the best answer.

1. 380 g = _____ kg
   A. 380,000
   B. 38,000
   C. 0.38
   D. 0.038

2. 8 L = _____ cL
   F. 8,000
   G. 800
   H. 0.08
   I. 0.008

3. 146 cm = _____ m
   A. 14,600
   B. 1,460
   C. 14.6
   D. 1.46

4. 32 cL + 480 mL + 89 cL = _____ cL
   F. 1,690 cL
   G. 601 cL
   H. 169 cL
   I. 125.8 cL

5. Which measurement is greatest?
   A. 3.6 km
   B. 36,000 m
   C. 360,000 cm
   D. 360,000 mm

PART 2 • Short Response

Record your answers in the space provided.

6. Sandy packs three gifts in a large box to mail to his cousins. The three gifts weigh 3 kg, 380 g, and 990 g. What is the total weight of the gifts, in kilograms?

_____

7. Lucy is framing a photograph in the shape of a rectangle. One side of a photograph is 20 cm long and the other side is 135 mm long. In centimeters, what is the length of the framing material needed for this photograph?

_____

PART 3 • Extended Response

Record your answer in the space provided.

8. Explain in your own words how the prefixes used in the metric system can help you convert between different units.

_____

_____

_____

Lesson 16.4

PART 1 • Multiple Choice

Choose the best answer.

1. LaTonya's father is building a fence for his garden. The garden is shaped like a rectangle, 40 feet by 20 feet. The fence will have posts at each corner and also at every 4 feet of fence. How many posts are needed?

   A. 15   C. 30

   B. 19   D. 34

2. Ginny is at the theater. She is looking for seat number 58. The first row has 10 seats, and the second row has 9 seats. The following rows continue to alternate between 10 seats and 9 seats. In which row is Ginny's seat?

   F. 5th row   H. 7th row

   G. 6th row   I. 8th row

3. Joanna is taking a walk. She begins by heading north. Then she turns to her left and walks. Later, she turns and walks back in the opposite direction. She makes a right turn, and then, makes another right turn. In which direction is she now facing?

   A. east   C. south

   B. north   D. west

4. A rectangular table is 12 feet long and 6 feet wide. Chairs for the table each need 3 feet of space for comfort. What is the greatest number of chairs that can fit around the table?

   F. 10   H. 14

   G. 12   I. 16

PART 2 • Short Response

Record your answers in the space provided.

5. A 3-mile trail has markers at its beginning, its end, and every $\frac{1}{2}$ mile. The trail leader decided that there should be markers every $\frac{1}{10}$ of a mile. How many more markers must be placed along the trail?

   _____

6. Marla surrounded her garden with square clay tiles. Each of the tiles measures 1 foot on each side. The garden is rectangular with a length of 10 feet and a width of 8 feet. She included tiles in the corner so that a complete path was formed. How many tiles did Marla use?

   _____

PART 3 • Extended Response

Record your answer in the space provided.

7. In a city, all the streets meet at right angles and are two-way streets. You can only make right turns, and U-turns are never allowed. Rajah is driving west on Roosevelt Avenue, but decides he wants to head east instead. Describe a way for Rajah to do this.

   _____

   _____

Lesson 16.5

PART 1 • Multiple Choice

Choose the best answer.

1. Select the statement that is true.

    A. ⁻4 < ⁻1   C. ⁻10 < ⁻12

    B. ⁻7 < ⁻9   D. ⁻6 > ⁻2

2. Select the statement that is true.

    F. 4 < ⁻2   H. 6 < ⁻24

    G. 3 < 2   I. 2 < 8

3. Select the statement that is true.

    A. ⁻7 > ⁻2   C. ⁻6 > ⁻3

    B. ⁻4 > ⁻1   D. ⁻3 > ⁻5

4. Select the statement that is true.

    F. ⁻23 > ⁻21

    G. ⁻34 > ⁻38

    H. ⁻51 > ⁻46

    I. ⁻102 > ⁻100

5. Select the statement that is true.

    A. ⁻15 < ⁻23

    B. ⁻11 > ⁻4

    C. ⁻9 < ⁻4

    D. ⁻26 > ⁻9

PART 2 • Short Response

Record your answers in the space provided.

6. Write an integer to represent a hill that is 298 feet above sea level.

    _____

7. Describe a situation that can be represented by ⁻9.

    _____

    _____

    _____

PART 3 • Extended Response

 Record your answer in the space provided.

8. Explain why ⁻6 < ⁻4.

    _____

    _____

    _____

Lesson 16.6

PART 1 • Multiple Choice

Choose the best answer.

1. Which of the following is the best estimate of the temperature in a movie theater?

   A. 30°F   C. 70°F

   B. 50°F   D. 90°F

2. Which of the following is the best estimate of the temperature of an ice cube?

   F. 0°C   H. 20°C

   G. 10°C   I. 30°C

3. The temperature outside is 20°C. Which is the best estimate of the equivalent Fahrenheit temperature?

   A. 70°F   C. 50°F

   B. 60°F   D. 40°F

4. The temperature of water in a teapot is 90°C. Which is the best estimate of the equivalent Fahrenheit temperature?

   F. 240°F   H. 170°F

   G. 200°F   I. 140°F

5. Tricia is going on vacation. The temperature will be about 30°C. Which of the following is the most appropriate item for Tricia to pack for her vacation?

   A. a hat   C. a winter coat

   B. a sweater   D. shorts

PART 2 • Short Response

Record your answers in the space provided.

6. At how many degrees Celsius does water boil?

   _____

7. At how many degrees Fahrenheit does water freeze?

   _____

PART 3 • Extended Response

Record your answer in the space provided.

8. If you know a Celsius temperature, you can estimate its equivalent Fahrenheit temperature using the rule "Multiply by 2 and add 30." Devise your own rule to estimate a Celsius temperature when a Fahrenheit temperature is known. Give an example of how your rule works.

   _____

   _____

   _____

Lesson 17.1

PART 1 • Multiple Choice

Choose the best answer.

1. There are *c* cars in a parking lot. Then 7 more cars drive into the lot. How many cars are now in the lot?

    A. 7   **C.** $c - 7$

    B. $7c$ **D.** $c + 7$

2. Diane has *f* oranges. She gives 3 of her oranges to Ben. How many oranges does Diane have now?

    F. 3   **H.** $3 - f$

    G. $f + 3$ **I.** $f - 3$

3. A carton of eggs contained 12 eggs. Jason used *n* eggs to make pancakes this morning. How many eggs were left in the carton?

    A. $12n$ **C.** $12 - n$

    B. $12 + n$ **D.** $n - 12$

4. What is the value of $b + 13$ for $b = 18$?

    F. 1,813   **H.** 21

    G. 31   **I.** 5

5. What is the value of $u - \frac{2}{3}$ for $u = 1\frac{1}{2}$?

    A. $\frac{2}{3}$ **C.** $1\frac{1}{6}$

    B. $\frac{5}{6}$ **D.** $2\frac{1}{3}$

PART 2 • Short Response

Record your answers in the space provided.

6. What is the value of $19.7 - s$ for $s = 12.1$?

    _____

7. What is the value of $5.8 + j$ for $j = 2.9$?

    _____

PART 3 • Extended Response

 Record your answer in the space provided.

8. Let *n* stand for the number of students in a class. Create a situation that might be described by the expression $n - 10$.

    _____

    _____

    _____

Lesson 17.2

PART 1 • Multiple Choice

Choose the best answer.

1. Roberto runs m miles each day. How many miles does he run in 5 days?

   A. $5 \times m$

   B. $5 + m$

   C. $\frac{m}{5}$

   D. $m \times m \times m \times m \times m$

2. Jan had 280 pennies. She divided the pennies equally among p piles. How many pennies are in each pile?

   F. $\frac{280}{p}$ H. $280 \times p$

   G. $\frac{p}{280}$ I. p

3. A box can hold 9 grapefruit. How many grapefruit can be packed into b boxes?

   A. $\frac{9}{b}$ C. $9 \times b$

   B. $\frac{b}{9}$ D. $9 + b$

4. What is the value of $12 \times d$ for $d = 7$?

   F. 19   H. 84

   G. 74   I. 127

5. What is the value of $\frac{72}{w}$ for $w = \frac{2}{3}$?

   A. $\frac{1}{108}$ C. 48

   B. $\frac{1}{48}$ D. 108

PART 2 • Short Response

Record your answers in the space provided.

6. What is the value of $\frac{k}{5}$ for $k = 45.5$?

   _____

7. What is the value of $\frac{1}{2} \times t$ for $t = 56$?

   _____

PART 3 • Extended Response

Record your answer in the space provided.

8. Let a stand for the number of teachers in a school. Create a situation that might be described by the expression $\frac{a}{4}$.

   _____

   _____

   _____

Lesson 17.3

PART 1 • Multiple Choice

Choose the best answer.

1. Evaluate 42   8   3.

    A. 102   C. 28

    B. 92   D. 18

2. Evaluate 10   6   2   1

    F. 16   H. 12

    G. 14   I. 7

3. What is the value of $5y$ 15   3 for y 10?

    A. 42   C. 32

    B. 38   D. 3

4. Which expression is equal to 50?

    F. $5 \times 5 + 5$ H. $(5 \times 5) + 5$

    G. $5 \times (5 + 5)$ I. $5 + 5 \times 5$

5. Which expression is NOT equal to 20?

    A. $(1 + 5)^2 - 4^2$

    B. $(2 + 3)(3 + 1)$

    C. $6 + 4 \times 2$

    D. $100 - 9^2 + 1^2$

PART 2 • Short Response

Record your answers in the space provided.

6. A teacher spent 10 minutes talking to a group of parents and 5 minutes speaking with each parent. There were 12 parents present. How many minutes in all did the teacher spend with the parents?

    _____

7. Rewrite the expression below with parentheses so that its value is 100.

    $$5 \times 5 + 8 + 7 \times 5$$

    _____

PART 3 • Extended Response

 Record your answer in the space provided.

8. Show the steps you take to evaluate the expression
$19 - 3^2 - 8 \div (2 + 2) + 5$
Explain how you used the order of operations.

    _____

    _____

Lesson 17.4

PART 1 • Multiple Choice

Choose the best answer.
An express delivery service charges $8, plus $4 per pound to deliver a package overnight. Use this information for problems 1–3.

1. How much would this service charge for a 2-pound package?

   A. $14   C. $20

   B. $16   D. $24

2. How much would this service charge for a 10-pound package?

   F. $40   H. $84

   G. $48   I. $120

3. Let w stand for the number of pounds that a package weighs and c stand for the cost, in dollars, charged by this delivery service. Which equation describes the situation correctly?

   A. $c = 8 + 4w$

   B. $c = 12w$

   C. $c = 4 + 8w$

   D. $c = 8 + 4 + w$

4. Ellen earns $8 per hour. If she works h hours in a week, which equation represents s, her total earnings, in dollars?

   F. $h = s + 8$

   G. $s = h + 8$

   H. $h = 8s$

   I. $s = 8h$

PART 2 • Short Response

Record your answers in the space provided.

5. The tables in a dining room can seat six people. In addition, there is one large table that can seat 15 people. How many people can be seated in this dining room if the large table and all of the 25 other tables are used?

   _____

6. An Internet Service Provider charges $10 per month plus $0.50 per hour for any hours over 100. Helene used the Internet for 126 hours last month. What was her charge?

   _____

PART 3 • Extended Response

 Record your answer in the space provided.

7. A car rental company charges $50 per day plus $0.25 per mile driven. Write an equation to describe this situation. Tell what each variable represents.

   _____

   _____

Lesson 17.5

PART 1 • Multiple Choice

Choose the best answer.
Use the coordinate
graph shown for
problems 1–3.

1. What are the coordinates of point *F*?

   A. (6, 1)   **C.** (1, 6)

   B. (5, 0)   **D.** (0, 5)

2. What are the coordinates of point *E*?

   F. (3, 4)   **H.** (4, 5)

   G. (4, 3)   **I.** (5, 4)

3. Which point has coordinates (1, 4)?

   A. Point *A*   **C.** Point *C*

   B. Point *B*   **D.** Point *D*

4. What is the value of *m* in the equation
   $m = 2n + 1$ if $n = 6$?

   F. 27   **H.** 13

   G. 14   **I.** 2.5

5. Which graph shows the function
   $y = x - 3$?

   A.   **C.**

   B. 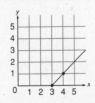   **D.**

PART 2 • Short Response

Record your answers in the space
provided.

6. What is the value of *y* in the
   equation $y = \frac{x}{4} + 9$ if
   $x = 24$?

   _____

7. A drive-in movie theater
   charges $6 per car plus $2
   per person in each car. Let
   p stand for the number of passengers
   in a car and *c* stand for the cost, in
   dollars, of going to this drive-in. Write
   an equation that relates *p* and *c*.

   _____

PART 3 • Extended Response

Record your answer in the space
provided.

8. Make a table for the function
   $y = 2x + 2$. Then graph the function.

   _____

   _____

Lesson 17.6

PART 1 • Multiple Choice

Choose the best answer. The graph below shows the number of greeting cards sold by Cards-R-Us daily for 6 days. Use for problems 1-6.

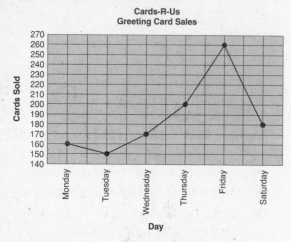

Cards-R-Us
Greeting Card Sales

1. What change occurred in sales between Monday and Tuesday?

   A. an increase by 5

   B. a decrease by 5

   C. a decrease by 10

   D. an increase by 10

2. What change occurred in sales between Tuesday and Wednesday?

   F. an increase by 10

   G. a decrease by 10

   H. a decrease by 20

   I. an increase by 20

3. Between which two days was the sharpest decrease in sales?

   A. Thursday and Friday

   B. Friday and Saturday

   C. Monday and Tuesday

   D. Wednesday and Thursday

PART 2 • Short Response

Record your answers in the space provided.

4. How many total greeting cards were sold over the 6 days?

   _____

5. If you owned the card store, on which day(s) of the week would you make sure you had the most help? Explain.

   _____

   _____

PART 3 • Extended Response

Record your answer in the space provided.

6. Explain why you think Friday has the most sales.

   _____

   _____

   _____

   _____

Lesson 18.1

PART 1 • Multiple Choice

Choose the best answer.

1. Solve for x: $x + 8 = 12$.

   A. $x = 1.5$

   B. $x = 4$

   C. $x = 20$

   D. $x = 96$

2. Solve for n: $17 + n = 24$.

   F. $n = 7$

   G. $n = 17$

   H. $n = 31$

   I. $n = 41$

3. Solve for w: $10 = w + 5$.

   A. $w = 50$

   B. $w = 15$

   C. $w = 5$

   D. $w = 2$

4. Solve for a: $16 = 4 + a$.

   F. $a = 64$

   G. $a = 20$

   H. $a = 12$

   I. $a = 4$

5. Solve for k: $33 = k + 19$.

   A. $k = 14$

   B. $k = 24$

   C. $k = 42$

   D. $k = 52$

PART 2 • Short Response

Record your answers in the space provided.

6. What value of h makes the equation $h + 9 = 22$ true?

   _____

7. What value of s makes the equation $41 = 14 + s$ true?

   _____

PART 3 • Extended Response

 Record your answer in the space provided.

8. A student solved the equation $x + 28 = 57$ and got the solution $x = 39$. Show how the student can see that the solution is incorrect by checking it in the original equation. Then find the correct solution.

   _____

   _____

Lesson 18.2

PART 1 • Multiple Choice

Choose the best answer.

1. Solve for y: $y + 11 = 20$.

   A. $y = \frac{11}{20}$ C. $y = 9$

   B. $y = \frac{20}{11}$ D. $y = 31$

2. Solve for d: $1\frac{4}{5} + d = 35\frac{3}{5}$.

   F. $d = 37\frac{2}{5}$ H. $d = 34\frac{4}{5}$

   G. $d = 36\frac{2}{5}$ I. $d = 33\frac{4}{5}$

3. Solve for p: $15.6 = p + 7.9$.

   A. $p = 7.7$ C. $p = 22.3$

   B. $p = 8.7$ D. $p = 23.3$

4. The road from Chicago to Minneapolis goes through Madison, Wisconsin. It is 409 miles from Chicago to Minneapolis, and 146 miles from Chicago to Madison. Which equation could be used to find m, the distance from Madison to Minneapolis?

   F. $m + 409 = 146$

   G. $m + 146 = 409$

   H. $m - 146 = 409$

   I. $m - 409 = 146$

5. An elevator can carry 1,000 pounds. Sid weighs 185 pounds. Which equation could be used to find f, the amount of freight Sid can bring on the elevator?

   A. $f - 185 = 1,000$

   B. $f - 1,000 = 185$

   C. $185 = f + 1,000$

   D. $1,000 = f + 185$

PART 2 • Short Response

Record your answers in the space provided.

6. What value of c makes the equation $12.8 + c = 50$ true?

   _____

7. What value of g makes the equation $73 = 39 + g$ true?

   _____

PART 3 • Extended Response

Record your answer in the space provided.

8. The gas tank in Malou's car can hold 14 gallons of gas. If Malou fills the tank by putting in 9.3 gallons of gas, how much gas was already in the tank? Write an equation and solve it. Show your steps.

   _____

   _____

   _____

Lesson 18.3

PART 1 • Multiple Choice

Choose the best answer.

1. Solve for q: $q - 18 = 35$.

   A. $q = 17$ **C.** $q = 43$

   B. $q = 27$ **D.** $q = 53$

2. Solve for x: $x - \frac{1}{3} = \frac{1}{6}$.

   F. $x = \frac{1}{2}$ **H.** $x = \frac{1}{6}$

   G. $x = \frac{1}{3}$ **I.** $x = \frac{2}{9}$

3. Solve for b: $9.24 = b - 3.77$.

   A. $b = 13.01$ **C.** $b = 12.01$

   B. $b = 12.91$ **D.** $b = 5.47$

4. Dan spent \$25 at a fair. He came home with \$48. Which equation could be used to find a, the number of dollars Dan brought to the fair?

   F. $a + 25 = 48$

   G. $a - 25 = 48$

   H. $25 - a = 48$

   I. $48 - a = 25$

5. After hiking 3.4 miles of a trail, Ellen had 6.8 miles left to go on the trail. Which equation could be used to find t, the length of the trail?

   A. $3.4 - t = 6.8$

   B. $6.8 - t = 3.4$

   C. $t - 3.4 = 6.8$

   D. $t + 3.4 = 6.8$

PART 2 • Short Response

Record your answers in the space provided.

6. What value of z makes the equation $32.1 = z - 15.5$ true?

   _____

7. What value of w makes the equation $w - \frac{1}{8} = \frac{5}{8}$ true? Write your answer as a fraction in simplest form.

   _____

PART 3 • Extended Response

Record your answer in the space provided.

8. Javy sells bottled water at a carnival. On Sunday, he sold 129 bottles and had 15 bottles left over at the end of the day. How many bottles did Javy have at the beginning of the day? Write a subtraction equation and solve it. Show your steps.

   _____

   _____

Lesson 18.4

PART 1 • Multiple Choice

Choose the best answer.

1. Solve for u: $u \times 6 = 216$.

   A. $u = 31$ C. $u = 210$

   B. $u = 36$ D. $u = 1{,}296$

2. Solve for a: $\frac{a}{6} = 16$.

   F. $a = 2$ H. $a = 24$

   G. $a = 8$ I. $a = 96$

3. Solve for g: $4 = g \div 2.5$.

   A. $g = 1.5$ C. $g = 6.5$

   B. $g = 1.6$ D. $g = 10$

4. For which equation is $n = 5$ a solution?

   F. $n + 2 = 3$

   G. $\frac{n}{2} = 10$

   H. $n - 2 = 7$

   I. $5n = 25$

5. Jon worked 40 hours last week and earned $380. Which equation could be used to find w, Jon's hourly wage?

   A. $\frac{w}{40} = 380$

   B. $\frac{w}{380} = 40$

   C. $40w = 380$

   D. $380w = 40$

PART 2 • Short Response

Record your answers in the space provided.

6. What value of j makes the equation $12j = 90$ true?

   _____

7. What value of v makes the equation $\frac{v}{15} = 3.2$ true?

   _____

PART 3 • Extended Response

Record your answer in the space provided.

8. At a children's museum, a 4-minute video runs continuously. If the museum is open for 8 hours each day, how many times can the video be shown in one day? Write an equation and solve it. Show your steps.

   _____

   _____

102 Use with Grade 5, Chapter 18, Lesson 4, pages 428-431.

Lesson 18.5

PART 1 • Multiple Choice

Choose the best answer. This graph shows the relationship between miles and kilometers. Use the graph for problems 1–3.

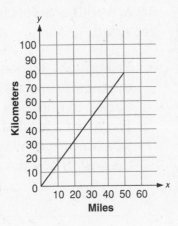

1. It is 44 mi from Bellemont to San Pedro. About how many km is that?

   A. 30   C. 70

   B. 60   D. 80

2. Pam rode her bicycle 15 mi to the beach. On the way back, she took a different route and rode 18 mi. About how many km did Pam ride in all?

   A. 65   C. 35

   B. 50   D. 20

3. A speed limit sign in Canada indicates that the speed limit is 65 km per hour. About many miles per hour is that?

   F. 35   H. 55

   G. 40   I. 100

This graph shows the relationship between centimeters and inches. Use the graph for problems 4–6.

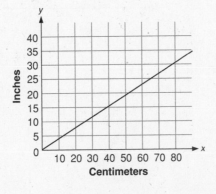

PART 2 • Short Response

Record your answers in the space provided.

4. A baby was 53 centimeters long at birth. To the nearest ten, how many inches long was this baby?

   _____

5. A yard is 36 inches long. To the nearest ten, how many centimeters is that?

   _____

PART 3 • Extended Response

Record your answer in the space provided.

6. Study the graph. Use it to describe a rule you can use to find about how many centimeters are equal to any number of inches. Use your rule to convert a height of 6 feet 2 inches to centimeters.

   _____

   _____

©Macmillan/McGraw-Hill

Lesson 18.6

PART 1 • Multiple Choice

Choose the best answer.

1. Solve for a: $4a + 2 = 14$.
   - A. $a = 48$ C. $a = 3$
   - B. $a = 4$ D. $a = 2$

2. Solve for d: $11 = 3d - 7$.
   - F. $d = 6$ H. $d = 12$
   - G. $d = 8$ I. $d = 54$

3. Solve for f: $\frac{f}{3} + 5 = 7$.
   - A. $f = 36$ C. $f = 5$
   - B. $f = 6$ D. $f = 4$

4. Solve for k: $\frac{5}{8} = \frac{k}{4} - \frac{1}{8}$.
   - F. $k = \frac{3}{16}$ H. $k = 2$
   - G. $k = 1\frac{1}{2}$ I. $k = 3$

5. Solve for g: $5g - 16 = {}^-6$.
   - A. $g = 50$ C. $g = 7$
   - B. $g = 15$ D. $g = 2$

PART 2 • Short Response

Record your answers in the space provided.

6. Ted bought 3 tickets to the theater and a $5 bucket of popcorn. If Ted spent a total of $26, how much was the cost of one ticket?

   _____

7. Darlene bought 12 apples and a newspaper for a total cost of $11.20. The newspaper cost $1. Write an equation that can be used to find the cost of an apple.

   _____

PART 3 • Extended Response

 Record your answer in the space provided.

8. Show the steps needed to solve the equation $11x + 23 = 56$.

   _____

   _____

   _____

   _____

Lesson 19.1

PART 1 • Multiple Choice

Choose the best answer.

1. What is the figure shown called?

   •————————•

   A. line
   B. line segment
   C. plane
   D. ray

2. Choose the words that complete the sentence correctly: A polygon must have at least _____.

   F. one side   H. three sides
   G. two sides   I. four sides

3. What is a polygon called if all its sides have equal length and all its angles have equal measure?

   A. congruent   C. regular
   B. normal   D. square

4. Which word describes the figures shown here?

   F. closed   H. horizontal
   G. diagonal   I. vertical

5. What figure has one endpoint?

   A. line
   B. line segment
   C. plane
   D. ray

PART 2 • Short Response

Record your answers in the space provided.

6. How many points are used to name a plane?

   _____

7. What is the point called where two sides of a polygon meet?

   _____

PART 3 • Extended Response

Record your answer in the space provided.

8. From the diagram below, name a point, ray, line segment, line, and plane.

   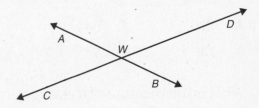

   _____

   _____

Lesson 19.2

PART 1 • Multiple Choice

Choose the best answer.

1. Which tool is used to measure angles?

   A. compass   C. ruler

   B. protractor   D. thermometer

2. Choose the words that complete the sentence correctly: An angle is formed by two _____ that have the same endpoint.

   F. lines

   G. line segments

   H. planes

   I. rays

3. What is the measure of the angle shown below?

   A. 116°   C. 64°

   B. 74°   D. 54°

4. What is the measure of the angle shown below?

   F. 190°   H. 110°

   G. 170°   I. 10°

5. What is the measure of the angle shown below?

   A. 32°   C. 54°

   B. 43°   D. 148°

PART 2 • Short Response

Record your answers in the space provided.

6. What is the measure, in degrees, of the angle shown below?

   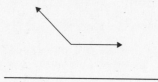

   _____

7. What is the measure of the angle shown below?

   _____

PART 3 • Extended Response

 Record your answer in the space provided.

8. Identify three different ways to name the angle shown below. Then find its measure.

   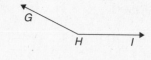

   _____

   _____

Lesson 19.3

PART 1 • Multiple Choice

Choose the best answer.

1. Which angle is obtuse?

    A.

    B.

    C.

    D.

2. What does the statement shown below mean?

    $$\overleftrightarrow{AB} \parallel \overleftrightarrow{MN}$$

    F. Line AB and line MN are vertical lines.

    G. Line AB is parallel to line MN.

    H. Line AB is perpendicular to line MN.

    I. Line AB intersects line MN.

Use this figure for problems 3–4.

3. Which lines are perpendicular?

    A. $\overleftrightarrow{GH}$ and $\overleftrightarrow{WD}$ C. $\overleftrightarrow{GW}$ and $\overleftrightarrow{HD}$

    B. $\overleftrightarrow{GH}$ and $\overleftrightarrow{HD}$ D. $\overleftrightarrow{GW}$ and $\overleftrightarrow{HW}$

4. Which lines are parallel?

    F. $\overleftrightarrow{WD}$ and $\overleftrightarrow{HD}$ H. $\overleftrightarrow{WD}$ and $\overleftrightarrow{GH}$

    G. $\overleftrightarrow{WD}$ and $\overleftrightarrow{HW}$ I. $\overleftrightarrow{WD}$ and $\overleftrightarrow{GW}$

PART 2 • Short Response

Record your answers in the space provided.

5. ∠A is a right angle. ∠B measures 163°. How many degrees greater is the measure of ∠B?

    _____

6. What type of angle is formed by the hands of a clock at 1:15?

    _____

PART 3 • Extended Response

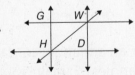

Record your answer in the space provided.

7. Can two lines intersect and form exactly one right angle? Explain your reasoning.

    _____

    _____

    _____

Lesson 19.4

PART 1 • Multiple Choice

Choose the best answer.

1. What type of triangle is shown here?

   A. equilateral   C. obtuse

   B. isosceles   D. scalene

2. What type of triangle is shown here?

   F. acute   H. right

   G. isosceles   I. scalene

3. What type of triangle is shown here?

   A. acute and equilateral

   B. acute and isosceles

   C. right and equilateral

   D. right and isosceles

4. A triangle contains a 105° angle and a 47° angle. What is the measure of its third angle?

   F. 28°   H. 48°

   G. 38°   I. 58°

5. A triangle contains a 40° angle and a 30° angle. What type of triangle is it?

   A. acute   C. right

   B. obtuse   D. isosceles

PART 2 • Short Response

Record your answers in the space provided.

6. What is the measure, in degrees, of ∠Y?

   _____

   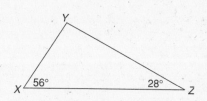

7. A triangle has sides of 8 cm, 12 cm and 16 cm. It contains an angle measuring more than 90° and two angles measuring less than 90°. Use two words from this lesson to describe the triangle.

   _____

PART 3 • Extended Response

 Record your answer in the space provided.

8. Can a triangle contain two obtuse angles? Explain your reasoning.

   _____

   _____

Lesson 19.5

PART 1 • Multiple Choice

Choose the best answer.

1. Which type of quadrilateral is shown?

   A. parallelogram

   B. rectangle

   C. rhombus

   D. trapezoid

2. Which type of quadrilateral is shown?

   F. parallelogram

   G. rectangle

   H. rhombus

   I. trapezoid

3. Which statement is true?

   A. All rectangles are rhombuses.

   B. All trapezoids are parallelograms.

   C. All rhombuses are parallelograms.

   D. All parallelograms are rectangles.

4. ABCD is a rhombus in which m∠B = 140° and m∠C = 40°. What is m∠A?

   F. 40°   H. 60°

   G. 50°   I. 140°

5. Which statement is false?

   A. All squares are parallelograms.

   B. All squares are rectangles.

   C. All rectangles are parallelograms.

   D. All rectangles are squares.

PART 2 • Short Response

Record your answers in the space provided.

6. Bob drew a quadrilateral that contained two 78° angles and one right angle. What was the measure, in degrees, of the fourth angle in Bob's quadrilateral?

   _____

7. Complete the paragraph below. A quadrilateral has four sides that each measure 90 cm. The quadrilateral must be a _____ and it could also be a _____.

   _____

PART 3 • Extended Response

Record your answer in the space provided.

8. Explain why every square is a rectangle but no trapezoid can be a rectangle.

   _____

   _____

   _____

Lesson 19.6

PART 1 • Multiple Choice

Choose the best answer.

1. A rectangular picture frame is 7 inches long and 5 inches wide. The frame is packed in a box whose base is square. What are the smallest possible dimensions for this box?

   A. 5 in. by 5 in.   **C.** 7 in. by 7 in.

   B. 6 in. by 6 in.   **D.** 12 in. by 12 in.

2. An ornament is shaped like an equilateral triangle that is 4 inches long on each side. The ornament is packed in a box whose base is square. What are the smallest possible dimensions for this box?

   F. 3 in. by 3 in.   **H.** 6 in. by 6 in.

   G. 4 in. by 4 in.   **I.** 12 in. by 12 in.

3. A tile is shaped like the trapezoid shown below. The tile is packed in a carton whose base is rectangular. What are the smallest possible dimensions for this carton?

   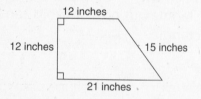

   A. 15 in. by 21 in.   **C.** 12 in. by 15 in.

   B. 12 in. by 21 in.   **D.** 12 in. by 12 in.

PART 2 • Short Response

Record your answers in the space provided.

4. A weight is shaped like the trapezoid in question 3. The weight is stored in a box whose base is square. What is the smallest possible side length, in inches, for this box?

   _____

5. Jorge orders a gift that comes in a box measuring 5 inches by 8 inches. He orders another gift that comes in a box measuring 6 inches by 6 inches. The two items are packed side-by-side in a carton whose base is rectangular. What are the smallest possible dimensions for this carton?

   _____

PART 3 • Extended Response

Record your answer in the space provided.

6. Desk lamps are sold in boxes with bases that measure 13 inches by 8 inches. Tom orders four lamps. Find the dimensions for the bases of two different cartons in which you could pack Tom's four lamps. Include diagrams with your work.

   _____

   _____

Lesson 20.1

PART 1 • Multiple Choice

Choose the best answer.
Use circle E below for problems 1–5.

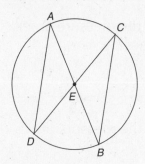

1. What is the center?

   A. $\overline{EA}$ C. $\overline{AEB}$

   B. $\overline{AB}$ D. Point *E*

2. Which of the following is a radius?

   F. $\overline{EA}$ H. $\overline{CB}$

   G. $\overline{AB}$ I. $\overline{DC}$

3. Which of the following is a diameter?

   A. $\overline{EA}$ C. $\overline{AD}$

   B. $\overline{AB}$ D. $\overline{EB}$

4. Which of the following is a chord?

   F. $\overline{EA}$ H. $\overline{CE}$

   G. $\overline{EB}$ I. $\overline{BC}$

5. Suppose m$\angle AED$ + m$\angle DEB$ + m$\angle BEC$ = 285°. What is m$\angle CEA$?

   A. 65°   C. 85°

   B. 75°   D. 95°

PART 2 • Short Response

Record your answers in the space provided.

6. The diameter of a circle is 21 feet. What is its radius, in feet?

   _____

7. What is the measure of the central angle formed by the hands of a clock at 9 A.M.?

   _____

PART 3 • Extended Response

 Record your answer in the space provided.

8. Explain how you would draw a circle whose diameter is 4 inches. Then draw your circle.

   _____

   _____

Lesson 20.2

PART 1 • Multiple Choice

Choose the best answer.
Use the figures shown below for problems 1–4.

1. Which triangles are congruent?

   A. 2 and 1   C. 2 and 4

   B. 2 and 3   D. 2 and 5

2. Which triangles are similar, but not congruent?

   F. 2 and 4   H. 1 and 3

   G. 2 and 5   I. 1 and 5

3. Which triangles are congruent?

   A. 1 and 2   C. 2 and 4

   B. 1 and 3   D. 2 and 3

4. Which triangles are neither similar nor congruent?

   F. 1 and 2   H. 3 and 4

   G. 1 and 3   I. 3 and 5

Quadrilaterals *ABCD* and *XYZW* are similar. Use the diagrams for problems 5–7.

5. Which side in *ABCD* corresponds to $\overline{YZ}$?

   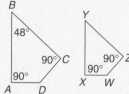

   A. $\overline{AB}$ C. $\overline{CD}$

   B. $\overline{BC}$ D. $\overline{DA}$

PART 2 • Short Response

Record your answers in the space provided.

6. What is the measure of ∠*Y*, in degrees?

   _____

7. What is the measure of ∠*W*?

   _____

PART 3 • Extended Response

 Record your answer in the space provided.

8. Two students drew rectangles with sides of length 6 inches and 8 inches each. Are the rectangles congruent? Explain.

   _____

   _____

   _____

Lesson 20.3

PART 1 • Multiple Choice

Choose the best answer.

1. What type of transformation is shown?

   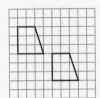

   A. glide reflection

   B. reflection

   C. rotation

   D. translation

2. What type of transformation is shown?

   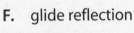

   F. glide reflection

   G. reflection

   H. rotation

   I. translation

3. The triangle with vertices (1, 1), (4, 1), and (3, 3) is transformed into the triangle with vertices (7,1), (4, 1), and (5, 3). What type of transformation was performed?

   A. glide reflection

   B. reflection

   C. rotation

   D. translation

4. Complete the sentence. A glide reflection is a combination of _____.

   F. a reflection and a translation

   G. a reflection and a rotation

   H. two reflections

   I. two translations

PART 2 • Short Response

Record your answers in the space provided.

5. How many degrees does the minute hand of a clock rotate in 30 minutes?

   _____

6. What combination of transformations is used in the diagram below?

   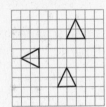

   _____

PART 3 • Extended Response

Record your answer in the space provided.

7. Find the letters of the alphabet that can be rotated 180° and look the same after rotation. Explain how you can show that these letters have this property.

   _____

   _____

   _____

Lesson 20.4

PART 1 • Multiple Choice

Choose the best answer.
This pattern is built using squares. Use the pattern for problems 1–2.

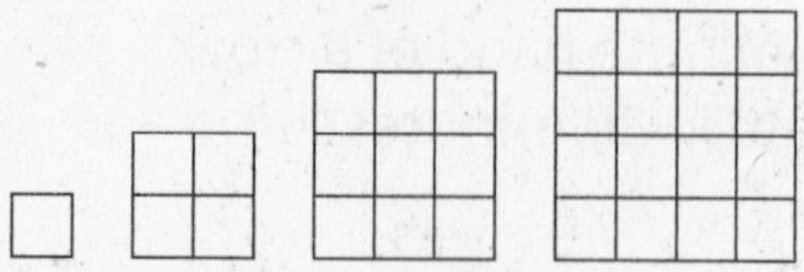

1. How many small squares will there be in the next figure of the pattern?

   A. 17 C. 27

   B. 25 D. 36

2. How many small squares will there be in the 10th figure of the pattern?

   F. 40 H. 80

   G. 50 I. 100

3. An artist uses a repeating pattern along a floor's tiles. The beginning of the pattern is shown below.

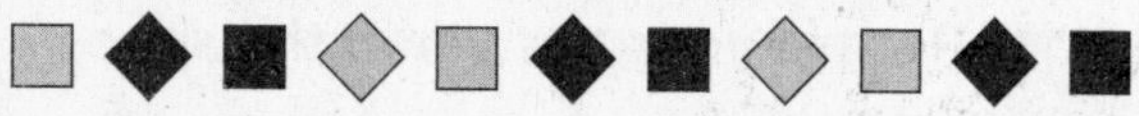

   What will the 30th tile look like?

   A.  C. 

   B.  D. 

This pattern is built using shaded triangles. Use it for problems 4–5.

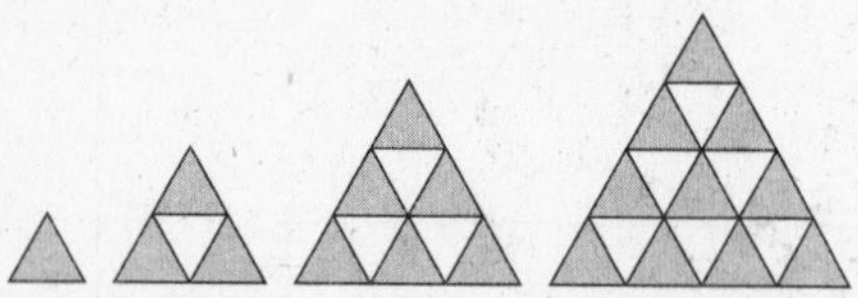

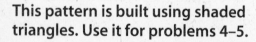

4. How many shaded triangles will there be in the next figure of the pattern?

   F. 15 H. 13

   G. 14 I. 12

PART 2 • Short Response

Record your answers in the space provided.

5. How many shaded triangles will there be in the 8th figure of the pattern?

   ____

6. How many unshaded triangles will there be in the 8th figure of the pattern?

   ____

PART 3 • Extended Response

THINK SOLVE EXPLAIN **Record your answer in the space provided.**

7. Write down the pattern for the number of shaded triangles. Then find the differences between the terms as you move along the pattern. Describe the pattern you see in those differences.

   ____

   ____

   ____

Lesson 21.1

PART 1 • Multiple Choice

Choose the best answer.

1. What is the perimeter of the rectangle shown?

   7 m

   5 m

   A. 12 m   C. 35 m

   B. 24 m   D. 70 m

2. What is the perimeter of the square shown?

   9 ft

   9 ft

   F. 81 ft   H. 36 ft

   G. 72 ft   I. 18 ft

3. On the grid shown, the sides of each small square have lengths of 1 unit. What is the perimeter of the rectangle drawn on the grid?

   A. 48 units   C. 32 units

   B. 36 units   D. 28 units

4. The polygon below is regular and its perimeter is 20 cm. What is the length of one side of the figure?

   F. 4 cm   H. 80 cm

   G. 5 cm   I. 100 cm

5. A vacant lot is shaped like a triangle. Two sides of the lot have lengths of 50 yards each, and the third side has a length of 60 yards. How many yards of fencing are needed to enclose the lot?

   A. 220   C. 160

   B. 170   D. 110

PART 2 • Short Response

Record your answers in the space provided.

6. The polygon below is regular and its perimeter is 100 m. What is the length of each side, in meters?

   _____

7. A park is shaped like a rectangle, 350 feet long and 185 feet wide. What is the perimeter of this park?

   _____

PART 3 • Extended Response

Record your answer in the space provided.

8. On the grid shown, the sides of each small square have lengths of 1 unit. Draw as many rectangles as you can in which each side is a whole number of units long and the perimeter is 16 units.

   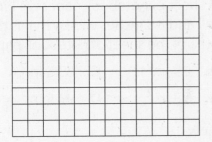

Lesson 21.2

PART 1 • Multiple Choice

Choose the best answer.

1. What is the area of the shaded rectangle?

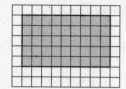

   A. 88 square units

   B. 50 square units

   C. 45 square units

   D. 28 square units

2. A square has sides of length 2 m. What is its area?

   F. 8 m   H. 4 m

   G. 8 m^2 I. 4 m^2

3. What is the area of the rectangle?

   A. 60 in.^{2} C. 34 in.^{2}

   B. 60 in.   D. 34 in.

4. The rectangle shown has an area of 68 square feet. What is its length?

   F. 30 feet   H. 12 feet

   G. 17 feet   I. 8 feet

5. A sheet of notebook paper is 11 inches tall and 8.5 inches wide. What is the area of the sheet of notebook paper?

   A. 39 in.^{2} C. 83.5 in.^{2}

   B. 78 in.^{2} D. 93.5 in.^{2}

PART 2 • Short Response

Record your answers in the space provided.

6. A rectangular lot has an area of 4,000 square feet. The length of the lot is 80 feet. What is the width, in feet?

   _____

7. Dee owns a square piece of land that is 100 feet long on each side. On this land, she built a house whose base is shaped like a rectangle 50 feet long and 30 feet wide. What is the area of the part of her land that surrounds the house?

   _____

PART 3 • Extended Response

Record your answer in the space provided.

8. The numbers 1, 4, 9, 16, 25, 36, . . . are called square numbers. Use what you learned in this lesson to explain why. Then find the next three square numbers.

   _____

   _____

Lesson 21.3

PART 1 • Multiple Choice

Choose the best answer.
A square has sides of length 12 cm. Use this square for problems 1–2.

1. If the side lengths are doubled, what happens to the square's perimeter?

   A. It becomes 12 cm greater.

   B. It becomes 24 cm greater.

   C. It becomes twice as great.

   D. It becomes four times as great.

2. If the side lengths are doubled, what happens to the square's area?

   F. It becomes 144 cm greater.

   G. It becomes twice as great.

   H. It becomes four times as great.

   I. It becomes sixteen times as great.

Two rectangles have sides 10 in. and 7 in. in length. The rectangles are joined along a shorter side of each to form a larger rectangle. Use this information for problems 3–4.

3. What is the area of the combined figure?

   A. 280 in.^{2} C. 70 in.^{2}

   B. 140 in.^{2} D. 68 in.^{2}

4. What is the perimeter of the combined figure?

   F. 54 in.   H. 61 in.

   G. 58 in.   I. 68 in.

PART 2 • Short Response

Record your answers in the space provided.

5. Two rectangles have sides 5 cm and 4 cm in length. The rectangles are joined along a longer side of each to form a larger rectangle. What is the perimeter, in centimeters, of the combined rectangle?

   _____

6. The length and width of a rectangle are doubled. The original rectangle had a length of 10 ft and a width of 4.5 ft. By how many square feet did the area increase?

   _____

PART 3 • Extended Response

Record your answer in the space provided.

7. What happens to the perimeter and area of a square if you triple the length of each of its sides? Use an example to help explain.

   _____

   _____

Lesson 21.4

PART 1 • Multiple Choice

Choose the best answer.
A construction company is building a new house. The house will be rectangular, measuring 50 feet long and 32 feet wide. The house will rest on a lot that measures 100 feet by 90 feet. Use this information for problems 1–3.

1. The builders will use string to mark off where the house will be. How much string will they use?

   A. 82 feet   **C.** 380 feet

   B. 164 feet   **D.** 1,600 feet

2. How much area will the house take up?

   F. 1,600 ft^2 **H.** 164 ft^2

   G. 1,500 ft^2 **I.** 82 ft^2

3. After the house is built, how much open area will be left on the lot?

   A. 216 ft^2 **C.** 8,400 ft^2

   B. 7,400 ft^2 **D.** 9,000 ft^2

4. Which problem can be solved by finding perimeter?

   F. You need to know how much paint to buy for a kitchen.

   G. You need to know how much framing material to buy to build a picture frame.

   H. You need to know how much material will cover a sofa.

   I. You need to know how much time it will take you to walk to school.

PART 2 • Short Response

Record your answers in the space provided.

5. A living room floor is rectangular with a length of 20 feet and a width of 12 feet. How many square feet of carpet will cover this floor?

   _____

6. Scott built a wood frame for a poster. The poster measures 22 inches by 17 inches. How much framing material did Scott use?

   _____

PART 3 • Extended Response

 Record your answer in the space provided.

7. Regina has a backyard vegetable garden. The garden is shaped like a rectangle 20 feet long and 40 feet wide. Describe a situation in which Regina would need to know the perimeter of her garden and a situation in which Regina would need to know the area of her garden.

   _____

   _____

   _____

Lesson 21.5

PART 1 • Multiple Choice

Choose the best answer.

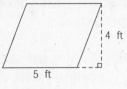

1. What is the area of the parallelogram?

   A. 20 ft^2 C. 18 ft^2

   B. 20 ft   D. 18 ft

2. What is the area of the parallelogram?

   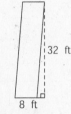

   F. 80 ft^2 H. 246 ft^2

   G. 160 ft^2 I. 256 ft^2

3. What is the area of the parallelogram?

   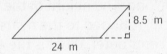

   A. 65 m^2 C. 204 m^2

   B. 102 m^2 D. 2,040 m^2

4. While riding a bus, José notices that one of its windows is shaped like a parallelogram. If the base of the parallelogram is 14 inches and its height is 9 inches, what is its area?

   F. 23 in.   H. 126 in.^{2}

   G. 46 in.   I. 252 in.^{2}

5. *ABCD* is a rhombus. What type of polygon is *ABXY*?

   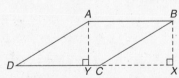

   A. rectangle   C. square

   B. rhombus   D. trapezoid

PART 2 • Short Response

Record your answers in the space provided.

6. The lot on which Kelly's house is built has the shape of a parallelogram. The length of the lot is 35 meters, and its width is 20 meters. What is the area of the lot, in square meters?

   _____

7. A diamond-shaped rug is actually shaped like a parallelogram. The base of the rug is 12 feet and the height is 8 feet. What is the area of a floor that this rug can cover?

   _____

PART 3 • Extended Response

 Record your answer in the space provided.

8. The sides of a parallelogram have lengths 11 cm and 5 cm. When an 11 cm side is used as base, the height is 4 cm. Draw this parallelogram and label its height and side lengths. Then find its area.

   _____

   _____

   _____

Lesson 22.1

PART 1 • Multiple Choice

Choose the best answer.

1. What is the area of the triangle?

   A. 9 in.   **C.** 18 in.

   B. 9 in.^{2} **D.** 18 in.^{2}

2. What is the area of the triangle?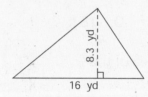

   F. 6 cm^2 **H.** 10 cm^2

   G. 7.5 cm^2 **I.** 12 cm^2

3. What is the area of the triangle?

   A. 16 in.^{2} **C.** 31.5 in.^{2}

   B. 27 in.^{2} **D.** 63 in.^{2}

4. A lot is shaped like a right triangle. Each of the perpendicular sides of the lot have length 200 feet. What is the area of the lot?

   F. 100 ft^2 **H.** 20,000 ft^2

   G. 2,000 ft^2 **I.** 40,000 ft^2

5. A canvas sail is shaped like a triangle. The base of the sail is 12 feet long. The height of the sail is 18 feet. How much material is in the sail?

   A. 30 ft^2 **C.** 98 ft^2

   B. 49 ft^2 **D.** 108 ft^2

PART 2 • Short Response

Record your answers in the space provided.

6. What is the area, in square yards, of the triangle?

   _____

7. A stamp is shaped like an equilateral triangle. Each side of the stamp is 3 cm long. The height of the stamp is 2.6 cm, rounded to the nearest tenth of a centimeter. What is the area of the stamp?

   _____

PART 3 • Extended Response

```
THINK
SOLVE
EXPLAIN
```
Record your answer in the space provided.

8. $\triangle ABC$ and $\triangle XBC$ share a base but are different triangles. Each has a height of 4 inches along the altitudes drawn to base $\overline{BC}$. Do the triangles have the same area or different areas? Explain your thinking.

   _____

   _____

Lesson 22.2

PART 1 • Multiple Choice

Choose the best answer.

1. What is the area of the trapezoid?

   A. 280 yd^2 **C.** 140 yd^2

   B. 160 yd^2 **D.** 100 yd^2

2. What is the area of the trapezoid?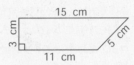

   F. 130 cm^2 **H.** 65 cm^2

   G. 78 cm^2 **I.** 39 cm^2

3. What is the area of the trapezoid?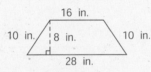

   A. 80 in.^{2} **C.** 176 in.^{2}

   B. 160 in.^{2} **D.** 352 in.^{2}

4. A car window is shaped like a trapezoid. Its bases are 22 inches and 18 inches long. Its height is 14 inches. What is the area of the window's surface?

   F. 140 in.^{2} **H.** 324 in.^{2}

   G. 280 in.^{2} **I.** 352 in.^{2}

5. The area of a trapezoid is 60 ft^2. Its bases have lengths of 9 feet and 21 feet. What is its height?

   A. 4 feet   **C.** 2 feet

   B. 3 feet   **D.** 1 foot

PART 2 • Short Response

Record your answers in the space provided.

6. A trapezoid has bases 3 m and 11 m in length. Its height is 2.4 m. What is its area, in square meters?

   _____

7. What is the area of the trapezoid?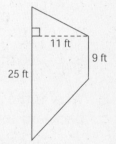

   _____

PART 3 • Extended Response

 Record your answer in the space provided.

8. Rewrite the formula for the area of a trapezoid using the Distributive Property. Explain how the new formula relates to triangles.

   _____

   _____

Lesson 22.3

PART 1 • Multiple Choice

Choose the best answer.

Janice's backyard is shaped like a rectangle 100 feet long and 60 feet wide. The yard is covered by grass except for a pool and its deck. The pool and its deck together are 40 feet long and 30 feet wide. The pool itself is 30 feet long and 20 feet wide. Use this information for problems 1-7.

1. What is the area of the pool and deck together?

   A. 600 ft^2 **C.** 2,400 ft^2

   B. 1,200 ft^2 **D.** 6,000 ft^2

2. What is the area of the portion of the backyard covered by grass?

   F. 1,200 ft^2 **H.** 5,400 ft^2

   G. 4,800 ft^2 **I.** 6,000 ft^2

3. What is the area of the deck, not counting the pool?

   A. 1,800 ft^2 **C.** 600 ft^2

   B. 1,200 ft^2 **D.** 300 ft^2

4. What fraction of the backyard is taken up by the pool and its deck?

   F. $\frac{1}{5}$ **H.** $\frac{1}{3}$

   G. $\frac{1}{4}$ **I.** $\frac{2}{5}$

5. Suppose the deck of the pool is the same width all the way around. How wide is it?

   A. 5 feet   **C.** 15 feet

   B. 10 feet   **D.** 20 feet

PART 2 • Short Response

Record your answers in the space provided.

A poster board is shaped like a rectangle 17 in. × 11 in. Kevin drew a picture on the poster board. It was shaped like a square with a triangle on top. The square had sides of length 5 in. The triangle had a base of 5 inches and a height of 4 in.

6. What is the area, in square inches, of Kevin's picture?

   _____

7. What is the area that remains on the poster board around the drawing?

   _____

PART 3 • Extended Response

Record your answer in the space provided.

8. A room has the dimensions shown below. Show two different ways to calculate its area.

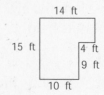

   _____

   _____

Lesson 22.4

PART 1 • Multiple Choice

Choose the best answer.

Each square of the grid has sides 1 inch long. Use the figure shown for problems 1 and 2.

1. What is the area of the shaded figure?

    A. 15 in.^{2} C. 17 in.^{2}

    B. 16 in.^{2} D. 18 in.^{2}

2. What is the perimeter of the shaded figure?

    F. 24 in.   H. 22 in.

    G. 23 in.   I. 21 in.

Each square of the grid has sides 1 centimeter long. Use the figure shown for problems 3–5.

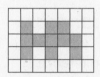

3. How many whole squares are shaded?

    A. 21   C. 28

    B. 26   D. 35

4. How many squares are partially shaded?

    F. 10   H. 8

    G. 9   I. 7

5. Which of the following is the best estimate of the area of the shaded region?

    A. 26 cm^2 C. 31 cm^2

    B. 28 cm^2 D. 36 cm^2

PART 2 • Short Response

Record your answers in the space provided.

Each square of the grid below has sides 1 inch long. Use the figure for problems 6 and 7.

6. What is the perimeter of the shaded region, in inches?

    _____

7. What is the area of the shaded region?

    _____

PART 3 • Extended Response

 Record your answer in the space provided.

8. Each square of the grid has sides 1 centimeter long. Estimate the area of the shaded region. Explain your method.

    _____

    _____

Lesson 22.5

PART 1 • Multiple Choice

Choose the best answer.
Note: For each problem, use $\pi \approx 3.14$.

1. What is the circumference of the circle, rounded to the nearest tenth?

    A. 28.2 cm   C. 56.5 cm

    B. 28.3 cm   D. 56.6 cm

2. What is the circumference of the circle, rounded to the nearest tenth?

    F. 44.0 in.   H. 22.0 in.

    G. 43.9 in.   I. 21.9 in.

3. What is the circumference of the circle, rounded to the nearest tenth?

    A. 18.8 ft   C. 37.7 ft

    B. 18.9 ft   D. 37.8 ft

4. A circular coin has radius of 8.2 mm. What is its circumference, rounded to the nearest tenth?

    F. 25.7 mm   H. 51.4 mm

    G. 25.8 mm   I. 51.5 mm

5. A pecan pie 40 feet in diameter was baked in Okmulgee, Oklahoma, in 1989. What was the pie's circumference, rounded to the nearest foot?

    A. 125 feet   C. 251 feet

    B. 126 feet   D. 252 feet

PART 2 • Short Response

Record your answers in the space provided.

6. What is the circumference of the circle, in meters, rounded to the nearest tenth?

    _____

7. A circular swimming pool has a diameter of 21 feet. What is the circumference of the pool? Round your answer to the nearest foot.

    _____

PART 3 • Extended Response

Record your answer in the space provided.

8. A circular track has a radius of 100 feet. Elena wants to run 2 miles. There are 5,280 feet in 1 mile. How many times should Elena run around the track? Show your method.

    _____

    _____

Lesson 23.1

PART 1 • Multiple Choice

Choose the best answer.

1. What 3-dimensional figure does the net make when cut and folded?

   A. cone   C. triangular prism

   B. cylinder   D. pyramid

2. What shape is the base of a cone?

   F. circle   H. square

   G. line   I. triangle

3. What 3-dimensional shape is commonly used for soup cans?

   A. cone   C. prism

   B. cylinder   D. pyramid

Use the figure below for problems 4–7.

4. What is the figure called?

   F. cone

   G. prism

   H. square pyramid

   I. triangular pyramid

5. How many vertices does it have?

   A. 3   C. 5

   B. 4   D. 6

PART 2 • Short Response

Record your answers in the space provided.

6. How many faces does the figure have?

   _____

7. How many edges does the figure have?

   _____

PART 3 • Extended Response

Record your answer in the space provided.

8. The figure shown is a pyramid whose base is a regular hexagon. Find how many faces, edges, and vertices it has. Then draw a net for the pyramid.

   _____

Lesson 23.2

PART 1 • Multiple Choice

Choose the best answer. Use the figure shown for problems 1–3.

1. Which is the front view?

    A.   C.

    B.   D.

2. Which is the side view?

    F.   H.

    G.   I.

3. Which is the top view?

    A.   C.

    B.   D.

4. What shape is the top view of a cylinder?

    F. circle   H. rectangle

    G. line   I. square

5. What shape is the side view of a cylinder?

    A. circle   C. rectangle

    B. line   D. square

PART 2 • Short Response

Record your answers in the space provided.

6.

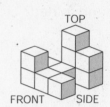

How many square faces appear in a top view of the figure shown here?

_____

7. What shape is the side view of a cone?

_____

PART 3 • Extended Response

 Record your answer in the space provided.

8. Draw a front view and a side view for the figure shown in problem 6.

©Macmillan/McGraw-Hill

Lesson 23.3

PART 1 • Multiple Choice

Choose the best answer.

1. What is the surface area of the rectangular prism?

   A. 92 ft²   **C.** 184 ft²

   B. 160 ft²   **D.** 224 ft²

2. What is the surface area of the rectangular prism?

   F. 504 yd²

   G. 494 yd²

   H. 252 yd²

   I. 192 yd²

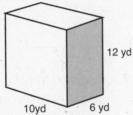

3. What is the surface area of the rectangular prism?

   A. 84 in.²

   B. 108 in.²

   C. 158 in.²

   D. 168 in.²

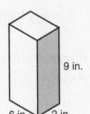

4. What is the surface area of the rectangular prism?

   F. 25.2 cm²   **H.** 53.4 cm²

   G. 26.7 cm²   **I.** 60.7 cm²

5. What is the surface area of the rectangular prism?

   A. 179 m²

   B. 358 m²

   C. 430 m²

   D. 458 m²

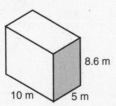

PART 2 • Short Response

Record your answers in the space provided.

6. What is the surface area, in square centimeters, of a cube that measures 6 cm on a side?

   _____

7. Explain how to find the surface area of a rectangular prism if you know the area of each face.

   _____

PART 3 • Extended Response

Record your answer in the space provided.

8. The length and width of a rectangular prism are each 2 m. The height is 8 m. Explain how you would find the surface area of the rectangular prism.

   _____

   _____

   _____

   _____

Lesson 23.4

PART 1 • Multiple Choice

Choose the best answer.

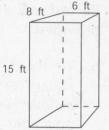

1. What is the volume of the rectangular prism?

   A. 720 ft³

   B. 720 ft²

   C. 516 ft³

   D. 516 ft²

2. What is the volume of the rectangular prism, rounded to the nearest tenth?

   F. 73.8 m³   **H.** 107.3 m³

   G. 73.9 m³   **I.** 107.4 m³

3. The base of a rectangular prism has an area of 32 cm². Its height is 5 cm. What is the volume of the prism?

   A. 37 cm³   **C.** 160 cm³

   B. 150 cm³   **D.** 5,120 cm³

4. Teri has an aquarium shaped like a rectangular prism. The aquarium is 10 inches tall, 25 inches wide, and 12 inches deep. What is the volume of the aquarium?

   F. 47 in.³   **H.** 3,000 in.³

   G. 1,340 in.³   **I.** 6,000 in.³

5. A rectangular box has volume of 72 ft³. Its length is 6 feet and its width is 3 feet. What is its height?

   A. 1 foot   **C.** 3 feet

   B. 2 feet   **D.** 4 feet

PART 2 • Short Response

Record your answers in the space provided.

6. The rectangular prism below has volume of 68 cm³. What is its length, in centimeters?

   _____

7. Each edge of a cube is 9 inches long. What is the volume of this cube?

   _____

PART 3 • Extended Response

Record your answer in the space provided.

8. A rectangular prism has a volume of 240 cm³. Find three different possible sets of values for its length, width, and height. Show your work.

   _____

   _____

Lesson 23.5

PART 1 • Multiple Choice

Choose the best answer.
Four identical cubes are arranged in a row to form a rectangular prism. Each of the cubes is 3 inches long. Use this information for problems 1–3.

1. What is the surface area of each cube?

   A. 36 in.^{2} C. 108 in.^{2}

   B. 54 in.^{2} D. 208 in.^{2}

2. What is the volume of the rectangular prism?

   F. 27 in.^{3} H. 162 in.^{3}

   G. 108 in.^{3} I. 208 in.^{3}

3. What is the surface area of the rectangular prism?

   A. 108 in.^{2} C. 162 in.^{2}

   B. 144 in.^{2} D. 208 in.^{2}

4. Prisms A and B are both 1 foot high. Prism A has a square base that is 8 feet long on each side. Prism B has a square base that is 5 feet long on each side. How much larger is the area of the base of Prism A than the area of the base of Prism B?

   F. 9 ft^2 H. 39 ft^2

   G. 29 ft^2 I. 49 ft^2

5. A shoe box measures 13 in. by 6 in. by 5 in. A box of pasta measures 8 in. by 5 in. by 3 in. The box of pasta is placed inside the shoe box. How much space inside the shoe box is NOT taken up by the box of pasta?

   A. 270 in.^{3} C. 195 in.^{3}

   B. 240 in.^{3} D. 120 in.^{3}

PART 2 • Short Response

Record your answers in the space provided.

6. A model consists of a rectangular prism on top of a cube. Each side of the cube measures 4 feet. The rectangular prism measures 3 feet by 3 feet by 2 feet. What is the volume of the model, in cubic feet?

   _____

7. Five cubes are stacked to form a tower. Each cube has sides that are 1 in. shorter than the cube beneath it. The largest cube has sides that are 5 in. long. What is the volume of the tower?

   _____

PART 3 • Extended Response

 Record your answer in the space provided.

8. Nine identical cubes are placed together to form a rectangular prism. Each cube has sides of length 1 in. The rectangular prism is 1 in. high, and its base is square. Then the cube in the center is removed from the figure. Draw the figure made from eight cubes. Then find its volume and surface area.

Lesson 24.1

PART 1 • Multiple Choice

Choose the best answer.

1. Which letter is symmetric about a line?

   A. F   C. M

   B. G   D. S

2. Which letter is NOT symmetric about a line?

   F. A   H. Y

   G. C   I. Z

3. How many lines of symmetry does the isosceles triangle have?

   A. 0   C. 2

   B. 1   D. 3

4. How many lines of symmetry does the parallelogram have?

   F. 0   H. 2

   G. 1   I. 4

5. Half of a polygon is shown below, along with its line of symmetry. If the full polygon is drawn, what is it?

   A. hexagon

   B. isosceles triangle

   C. parallelogram

   D. trapezoid

PART 2 • Short Response

Record your answers in the space provided.

6. How many lines of symmetry does the regular octagon shown below have?

   _____

7. How many lines of symmetry does this figure have?

   _____

PART 3 • Extended Response

 Record your answer in the space provided.

8. Draw a shape that has exactly two lines of symmetry. Include those lines on your drawing, and explain how you can tell that each is a line of symmetry.

   _____

   _____

Lesson 24.2

PART 1 • Multiple Choice

Choose the best answer.

1. Which letter has rotational symmetry?

   A. M   C. T

   B. N   D. U

2. Which number has rotational symmetry?

   F. 3   H. 7

   G. 5   I. 8

3. The figure shown has rotational symmetry. What is the smallest fraction of a full turn needed for it to look the same?

   A. $\frac{1}{6}$ C. $\frac{1}{3}$

   B. $\frac{1}{4}$ D. $\frac{1}{2}$

4. The figure shown has rotational symmetry. What is the smallest fraction of a full turn needed for it to look the same?

   F. $\frac{1}{6}$ H. $\frac{1}{3}$

   G. $\frac{1}{4}$ I. $\frac{1}{2}$

5. Which figure must have rotational symmetry?

   A. acute triangle

   B. isosceles triangle

   C. rectangle

   D. trapezoid

PART 2 • Short Response

Record your answers in the space provided.

6. The figure below has rotational symmetry. What is the smallest number of rotated degrees needed for it to look the same?

   _____

7. The figure shown has rotational symmetry. What is the smallest number of rotated degrees needed for it to look the same?

   _____

PART 3 • Extended Response

Record your answer in the space provided.

8. Draw two shapes of your own that have rotational symmetry and that look the same after less than $\frac{1}{2}$ turn. Explain how you can tell.

   _____

   _____

Lesson 24.3

PART 1 • Multiple Choice

Choose the best answer.

1. One person can sit on each side of a square table. Five tables are put in a row to form one rectangular table. How many can sit at this table?

   A. 12   C. 16

   B. 14   D. 20

2. One person can sit on each side of a square table. Nine tables are pushed together to form one square table. How many can sit at this table?

   F. 12   H. 20

   G. 16   I. 36

Allentown is 150 mi north of Taris, which is 50 mi west of Kelina. Kelina is 200 mi south of Brackenville, which is 50 mi east of Corsio. Corsio is 100 mi south of Preston. Use for problems 3–5.

3. Which town is 50 miles directly north of Allentown?

   A. Brackenville   C. Kelina

   B. Corsio   D. Preston

4. Which is 300 miles north of Taris?

   F. Allentown   H. Corsio

   G. Brackenville   I. Preston

5. How far is Allentown from Brackenville?

   A. 50 miles east and 50 miles south

   B. 50 miles east and 50 miles north

   C. 50 miles west and 50 miles south

   D. 50 miles west and 50 miles north

PART 2 • Short Response

Record your answers in the space provided.

6. The numbers on the opposite faces of a number cube have a sum of 7. When a number cube is tossed, the number 4 appears on the top face. What is the sum of the other numbers that are visible on the cube?

_____

7. A carton's base is 20 inches × 31 inches. Boxes whose bases are shaped like squares 3 inches on a side must be packed into the carton. How many boxes can fit in one layer within the carton?

_____

PART 3 • Extended Response

 Record your answer in the space provided.

8. Alan has 16 pieces of toy fencing. Each piece is 1 foot long. He wants to enclose a rectangular pen with his pieces. Find the dimensions of the pen with the largest area that Alan can enclose. Show your work.

_____

_____

©Macmillan/McGraw-Hill

Lesson 24.4

PART 1 • Multiple Choice

Choose the best answer.

1. Which shape tessellates?

   A.   C. ▢

   B. ○   D. ✦

2. Which shape tessellates?

   F. ◺   H. ◗

   G. ✺   I. ⬠

3. Which shape tessellates?

   A. ⬭   C.

   B. △   D. ⌒

4. Which shape tessellates?

   F. ☽   H. ☆

   G.   I. ◇

5. Which shape tessellates?

   A.   C. ▭

   B. ▱   D.

PART 2 • Short Response

Record your answers in the space provided.

6. A regular hexagon will tessellate. How many sides does a regular hexagon have?

   _____

7. Show how the trapezoid below tessellates.

   _____

PART 3 • Extended Response

THINK SOLVE EXPLAIN Record your answer in the space provided.

8. Create a shape that tessellates and then show that it tessellates.

   _____

   _____

   _____

   _____

   _____

Lesson 25.1

PART 1 • Multiple Choice

Choose the best answer. Use the picture of the circles below for problems 1-3.

1. What is the ratio of shaded circles to un-shaded circles?

   A. 8 to 4   **C.** 4 to 8

   B. 8 to 12   **D.** 4 to 12

2. What is the ratio of shaded circles to the total number of circles?

   F. $\frac{4}{12}$ **H.** $\frac{8}{12}$

   G. $\frac{4}{8}$ **I.** $\frac{12}{8}$

3. What is the ratio of un-shaded circles to the total number of circles?

   A. 12:4   **C.** 4:8

   B. 8:4   **D.** 4:12

Use the picture of the figures below for problems 4-7.

4. What is the ratio of stars to the total number of objects?

   F. 3 to 22   **H.** 6 to 19

   G. 6 to 22   **I.** 6 to 8

5. What is the ratio of squares to circles?

   A. $\frac{8}{22}$ **C.** $\frac{8}{6}$

   B. $\frac{8}{14}$ **D.** $\frac{8}{5}$

PART 2 • Short Response

Record your answers in the space provided.

6. What is the denominator of the ratio of triangles to stars?

   _____

7. Write the ratio of circles to triangles in all three ways.

   _____

   _____

   _____

PART 3 • Extended Response

Record your answer in the space provided.

8. The ratio of boys to girls in a classroom is 13 to 12. What is the ratio of boys to the total number of students in the classroom? Explain.

   _____

   _____

   _____

   _____

Lesson 25.2

PART 1 • Multiple Choice

Choose the best answer.

1. Which ratio is equivalent to 4:5?

   A. 5:6   C. 12:15

   B. 8:9   D. 44:54

2. Which ratio is NOT equivalent to $\frac{6}{16}$?

   F. $\frac{3}{8}$

   G. $\frac{15}{40}$

   H. $\frac{24}{54}$

   I. $\frac{60}{160}$

3. What value of n makes the number sentence below true?

   $8:1 = n:9$

   A. $n = 72$ C. $n = 17$

   B. $n = 63$ D. $n = 16$

4. There are 18 workers at a travel agency. Twelve work part-time and six work full-time. What is the ratio of full-time to part-time workers?

   F. 1:2   H. 2:1

   G. 1:3   I. 3:1

5. Which pair of ratios is equivalent?

   A. $\frac{4}{5}$ and $\frac{5}{6}$

   B. $\frac{3}{4}$ and $\frac{9}{10}$

   C. $\frac{2}{3}$ and $\frac{22}{32}$

   D. $\frac{1}{2}$ and $\frac{13}{26}$

PART 2 • Short Response

Record your answers in the space provided.

6. What value of n makes the number sentence below true?

   $35:21 = n:3$

   _____

7. A recipe uses sugar and cocoa in a ratio of 5:3. Helene made this recipe using 15 tablespoons of sugar. How many tablespoons of cocoa did Helene use?

   _____

PART 3 • Extended Response

 Record your answer in the space provided.

8. It rained on 8 of the 30 days in April and 9 of the 31 days in May. Wim claimed that the ratio of rainy days to days on which it didn't rain was the same in both months. Do you agree? Explain your thinking.

   _____

   _____

Lesson 25.3

PART 1 • Multiple Choice

Choose the best answer.

1. What is the missing value needed to make the rates equal?

   30 mi in 4 h = ☐ mi in 12 h

   A. 240   C. 110

   B. 120   D. 90

2. What is the missing value needed to make the rates equal?

   12 pages in 8 min = ☐ pages in 40 min

   F. 384   H. 44

   G. 60   I. 27

3. What is the missing value to make the rates equal?

   15 for $35.25 = 75 for ☐

   A. $176.25   C. $95.25

   B. $175.25   D. $40.25

4. What is the missing value of the unit rate?

   16 for $64 = 1 for ☐

   F. $16   H. $4

   G. $8   I. $2

5. What is the missing value of the unit rate?

   390 mi on 13 gal = ☐ mi on 1 gal

   A. 32.6   C. 21

   B. 30   D. 3

PART 2 • Short Response

Record your answers in the space provided.

6. Bob can type 55 words in 3 minutes. At this rate, how many words can he type in 24 minutes?

   _____

7. Grace drove 220 miles on 10 gallons of gas. Denise drove 125 miles on 5 gallons of gas. Find the unit rate for each and compare.

   _____

   _____

   _____

PART 3 • Extended Response

Record your answer in the space provided.

8. Kirk can read 8 pages in 5 minutes. If he starts reading at 7:15 PM, at what time should he be done reading 240 pages? Explain.

   _____

   _____

   _____

Lesson 25.4

PART 1 • Multiple Choice

Choose the best answer.

1. A package of 8 juice boxes costs $1.89. What is the unit price, rounded to the nearest cent?

    A. $0.21   C. $0.23

    B. $0.22   D. $0.24

2. A 32-ounce bottle of olive oil costs $5.25. What is the unit cost, rounded to the nearest cent?

    F. $0.16   H. $0.19

    G. $0.17   I. $0.20

3. Todd wants to buy limes. Which is the best buy?

    A. 5 limes for $1

    B. 12 limes for $3

    C. 4 limes for $0.89

    D. 8 limes for $2

4. Wendy wants to buy ears of corn. Which is the best buy?

    F. 1 ear for $0.35

    G. 3 ears for $1

    H. 10 ears for $3

    I. 12 ears for $4

5. Cellophane tape is sold in rolls of different length. Which is the best buy?

    A. 250 inches for $0.99

    B. 500 inches for $1.99

    C. 750 inches for $2.69

    D. 900 inches for $3.39

PART 2 • Short Response

Record your answers in the space provided.

6. A box of 12 golf balls costs $21.89. What is the unit cost, rounded to the nearest cent?

    _____

7. A season ticket to a professional basketball team costs $2,200. The team plays 40 games. What is the unit cost?

    _____

PART 3 • Extended Response

Record your answer in the space provided.

8. A 32-ounce container of milk costs $0.89. A 128-ounce container of milk costs $2.99. Show how you can tell which is the better buy. Describe a situation where you would choose the container that is *not* the better buy.

    _____

    _____

Lesson 25.5

PART 1 • Multiple Choice

Choose the best answer.
The table shows how long it takes Marisa and David to do chores. Use the table for problems 1–3.

Chore	Time
Take out trash	3 min
Clean bathroom	45 min
Sort recycling	20 min
Vacuum	1 h
Unload dishwasher	10 min
Weed	2 h

1. On Saturdays, Marisa cleans both bathrooms in her house. Which is the most reasonable estimate of how long it will take her to do this?

   A. 1 hour   C. 2 hours

   B. $1\frac{1}{2}$ hours   D. $2\frac{1}{2}$ hours

2. While Marisa vacuums on Saturday, David unloads the dishwasher, sorts the recycling, and takes out the trash. Which statement is most reasonable?

   F. It will take David about half as long as Marisa.

   G. It will take David almost as long as Marisa.

   H. It will take David a few minutes longer than Marisa.

   I. It will take David about twice as long as Marisa.

3. David weeds the garden once in April, then twice in June, once in July, twice in August, and once in October. Which is the most reasonable estimate of the amount of time he spends weeding each year?

   A. 7 hours   C. 12 hours

   B. 10 hours   D. 14 hours

PART 2 • Short Response

Record your answers in the space provided.
It costs $6 to wash a car and $15 to wash and polish a car. Use this information for problems 4–5.

4. One Saturday, 28 cars were washed and polished. The total amount earned that day was $864. How many cars were only washed?

   _____

5. One Sunday, 89 cars were only washed and 31 cars were washed and polished. What was the total amount earned that day?

   _____

PART 3 • Extended Response

 Record your answer in the space provided.

6. Emily earns $24 per hour. She works 40 hours per week, on average. Make a reasonable estimate of the amount she will earn in a year. Show your method.

   _____

   _____

Lesson 25.6

PART 1 • Multiple Choice

Choose the best answer.

1. If an actual room that is 20 feet wide measures 5 inches wide on a drawing, what is the scale?

   A. 1 in. 15 ft   **C.** 1 in. 5 ft

   B. 1 in. 8 ft   **D.** 1 in. 4 ft

2. If an actual room that is 20 meters wide measures 10 centimeters on a drawing, what is the scale?

   F. 1 cm 10 m   **H.** 1 cm 2 m

   G. 1 cm 5 m   **I.** 1 cm 1 m

3. The scale on a map is 1 inch: 50 miles. If the map distance between two cities is 5 inches, what is the actual distance between the cities?

   A. 250 mi   **C.** 200 mi

   B. 225 mi   **D.** 100 mi

4. The scale on a map is 1 centimeter: 80 kilometers. If the map distance between two cities is 6 centimeters, what is the actual distance between the cities?

   F. 400 km   **H.** 520 km

   G. 480 km   **I.** 540 km

5. The scale on a map is 1 centimeter: 120 kilometers. If the map distance between two cities is 9 centimeters, what is the actual distance between the cities?

   A. 744 km   **C.** 1,080 km

   B. 860 km   **D.** 2,160 km

PART 2 • Short Response

Record your answers in the space provided.

6. The distance between two cities is 450 miles. The scale on the map is 1 inch: 50 miles. What is the distance between the cities on the map, in inches?

   _____

7. A room that is 15 feet by 18 feet is drawn using a scale of 1 inch: 3 feet. Will the drawing fit on a piece of paper $8\frac{1}{2}$ in. by 11 in.? Explain.

   _____

   _____

PART 3 • Extended Response

Record your answer in the space provided.

8. An architect wants to draw a scale drawing of a house that is 40 feet by 80 feet. He wants to use a piece of paper 11 inches by 14 inches. Select a scale that would fit on the paper. Justify your scale selection.

   _____

   _____

   _____

Lesson 26.1

PART 1 • Multiple Choice

Choose the best answer.

1. When an outcome probably will not happen, it is called _____

   A. likely   C. probable

   B. most likely   D. unlikely

Use the spinner shown below for problems 2–5.

2. If the spinner is spun once, how many possible outcomes are there?

   F. 1   H. 4

   G. 3   I. 7

3. If the spinner is spun once, which outcome is most likely to occur?

   A. 1   C. 5

   B. 3   D. 7

4. If the spinner is spun once, which outcome is least likely to occur?

   F. 1   H. 5

   G. 3   I. 7

5. Suppose one of the sections labeled 5 is relabeled as 7 and one of the sections labeled 1 is relabeled as 3. If the new spinner is spun once, which outcome is most likely to occur?

   A. 1   C. 5

   B. 3   D. 7

PART 2 • Short Response

Record your answers in the space provided. Use the spinner below for problems 6 and 7.

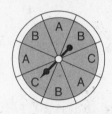

6. How many different outcomes are possible on this spinner?

   _____

7. Which outcome is least likely to occur if the spinner is spun once?

   _____

PART 3 • Extended Response

Record your answer in the space provided.

8. Design a spinner that has 12 equal sections, using the colors red, blue, yellow, and green. On your spinner, yellow should be the most likely color to be spun and green should be least likely.

Lesson 26.2

PART 1 • Multiple Choice

Choose the best answer.
Use the spinner below for
problems 1-5.

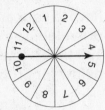

1. If you spin the spinner,
   what is $P(5)$?

   A. $\frac{1}{17}$ C. $\frac{1}{5}$

   B. $\frac{1}{12}$ D. $\frac{5}{12}$

2. If you spin the spinner, what is
   $P(2 \text{ or } 6)$?

   F. $\frac{1}{2}$ H. $\frac{1}{6}$

   G. $\frac{1}{3}$ I. $\frac{1}{12}$

3. If you spin the spinner, what is $P(\text{even number})$?

   A. $\frac{1}{2}$ C. $\frac{1}{4}$

   B. $\frac{1}{3}$ D. $\frac{1}{8}$

4. If you spin the spinner, what is
   $P(\text{a number less than 13})$?

   F. 0   H. $\frac{3}{4}$

   G. $\frac{1}{2}$ I. 1

5. Which statement is true?

   A. Spinning a 5 is more likely than
      spinning a 12.

   B. Spinning a 5 is equally as likely as
      spinning a 12.

   C. Spinning a 5 is less likely than
      spinning a 12.

   D. Spinning a 5 is impossible.

PART 2 • Short Response

Record your answers in the space
provided. Use the spinner
below for problems 6-8.

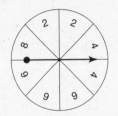

6. If you spin the spinner,
   which number is less likely
   to occur than 2?

   _____

7. How does the probability of spinning
   a 6 compare to that of spinning a 2?
   Explain.

   _____

   _____

PART 3 • Extended Response

Record your answer in the space
provided.

8. What is the probability of spinning a
   multiple of 2? Explain the meaning of
   this probability.

   _____

   _____

Lesson 26.3

PART 1 • Multiple Choice

Choose the best answer. A number cube marked 1 to 6 is tossed 150 times. Use this for problems 1–4.

1. What is the theoretical probability of tossing a number greater than 2?

   A. $\frac{1}{6}$ C. $\frac{1}{2}$

   B. $\frac{1}{3}$ D. $\frac{2}{3}$

2. Which is the most reasonable prediction of the number of times 3 will be tossed?

   F. 15   H. 25

   G. 20   I. 30

3. Which is the most reasonable prediction of the number of times an odd number will be tossed?

   A. 75   C. 50

   B. 60   D. 30

4. Suppose the number 6 was tossed 38 times. What can you conclude?

   F. The experimental probability of tossing 6 is greater than the theoretical probability of tossing 6.

   G. The experimental probability of tossing 6 is greater than the experimental probability of tossing 5.

   H. The experimental probability of tossing 6 is less than the theoretical probability of tossing 6.

   I. The experimental probability of tossing 6 is less than the experimental probability of tossing 5.

PART 2 • Short Response

Record your answers in the space provided. The spinner is spun 120 times. Use the spinner below for problems 5 and 6.

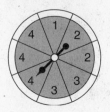

5. What is the most reasonable prediction of the number of times the spinner will land on a section labeled 4?

_____

6. What is the theoretical probability of the spinner landing on a section labeled 3?

_____

PART 3 • Extended Response

Record your answer in the space provided.

7. A number cube marked 1 to 6 is tossed 30 times. Dana says, "Each number must be tossed 5 times." Is this true? Explain your thinking.

_____

_____

PART 1 • Multiple Choice

Choose the best answer.
Quentin wants to know what type of socks students wear to school. He asked students as they entered school one day to identify their sock color. The table shows his findings. Use the table for problems 1–3.

Sock Color	Black	Blue	Red	White
Number of Students	12	11	6	61

1. How many students answered Quentin's question?

   A. 4
   B. 60
   C. 80
   D. 90

2. Which color sock was worn by the greatest number of students?

   F. Black
   G. Blue
   H. Red
   I. White

3. Suppose you had to predict a student's sock color. Which color is least likely?

   A. Black
   B. Blue
   C. Red
   D. White

Rosita wanted to predict how a toy car would land if dropped to the floor. She dropped the car a number of times and recorded the results shown below. Use the table for problems 4–6.

Toy Car Lands On	Wheels	Top	Left Side	Right Side
Number of Times	5	21	14	10

4. How many times did Rosita conduct her experiment?

   F. 60
   G. 50
   H. 40
   I. 4

PART 2 • Short Response

Record your answers in the space provided.

5. Based on Rosita's experiment, what is the probability that the toy car will land on its top?

   _____

6. Based on Rosita's experiment, what is the probability that the toy car will land on its right side?

   _____

PART 3 • Extended Response

 Record your answer in the space provided.

7. The United States Mint is issuing commemorative quarters for each of the 50 states. If you get a quarter in change at a store, what is the probability that it is one of the state commemorative quarters? Design an experiment you could use to estimate this probability.

   _____

   _____

   _____

Lesson 26.5

PART 1 • Multiple Choice

Choose the best answer.
A quarter and a nickel are tossed together. Use this information for problems 1 and 2.

1. What is the probability of tossing tails on both coins?

   A. $\frac{1}{8}$ C. $\frac{1}{3}$

   B. $\frac{1}{4}$ D. $\frac{1}{2}$

2. Suppose the coins are tossed together 40 times. What is the best prediction of the number of times both coins will show tails?

   F. 10   H. 20

   G. 15   I. 30

Use the two spinners below for problems 3–6.

3. Suppose the second spinner is spun once. What is the probability of spinning an odd number?

   A. $\frac{1}{2}$ C. $\frac{2}{3}$

   B. $\frac{1}{3}$ D. $\frac{3}{5}$

4. Suppose both spinners are spun once. What is the probability that both will land on an even number?

   F. $\frac{2}{5}$ H. $\frac{5}{6}$

   G. $\frac{1}{5}$ I. $\frac{1}{6}$

PART 2 • Short Response

Record your answers in the space provided.

5. Suppose both spinners are spun 60 times. What is the best prediction of the number of times both spinners will land on an even number?

   _____

6. Suppose both spinners are spun once. What is the probability that the first spinner will land on an even number and the second spinner will land on an odd number?

   _____

PART 3 • Extended Response

 Record your answer in the space provided.

7. Two 1-to-6 number cubes are tossed. Draw a tree diagram to show the possible outcomes. Use your tree diagram to find the probability that the same number is tossed on each cube.

   _____

   _____

Lesson 26.6

PART 1 • Multiple Choice

Choose the best answer.

1. A sandwich shop has 5 different types of sandwiches and 6 different types of chips to choose from. If you choose a sandwich and bag of chips, what is the total number of possible outcomes?

   A. 6   C. 30

   B. 11   D. 60

2. What is the number of possible outcomes if you toss a coin and toss a 1-to-6 number cube?

   F. 18   H. 6

   G. 12   I. 2

Use the spinners for problems 3-4.

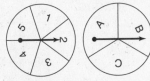

3. What is the total number of possible outcomes if you spin each spinner once?

   A. 2   C. 8

   B. 5   D. 15

4. What is the probability of spinning 4 and A on one spin of the spinners?

   F. $\frac{1}{15}$ H. $\frac{1}{5}$

   G. $\frac{1}{8}$ I. $\frac{1}{2}$

Use the spinner for problems 5-8.

5. What is the total number of possible outcomes for spinning the spinner twice?

   A. 10   C. 100

   B. 20   D. 1,000

PART 2 • Short Response

Record your answers in the space provided.

6. If you spun the spinner twice and added the sum of the spins, what is the largest sum you could get?

   _____

7. If you spun the spinner twice and added the sum of the spins, how many ways could you get a sum of 4? Explain.

   _____

PART 3 • Extended Response

THINK
SOLVE
EXPLAIN

Record your answer in the space provided.

8. If you spun the spinner twice and added the sum of the spins, what are all of the possible sums you could get? Create a table to justify your answer.

   _____

Lesson 27.1

PART 1 • Multiple Choice

Choose the best answer.

1. How is 0.40 written as a percent?

   A. 0.4%   C. 40%

   B. 4%   D. 400%

2. How is $\frac{6}{100}$ written as a percent?

   F. 60%   H. 0.6%

   G. 6%   I. 0.06%

3. How is 50:100 written as a percent?

   A. 50,000%

   B. 5,000%

   C. 500%

   D. 50%

4. In a bag of 5 apples, 3 are red and 2 are yellow. What percent of the apples are red?

   F. 6%

   G. 15%

   H. 60%

   I. 150%

5. In a survey of 20 students, 17 were in favor of holding a spring carnival. What percent of the students is this?

   A. 1,700%

   B. 85%

   C. 17%

   D. 8.5%

PART 2 • Short Response

Record your answers in the space provided.

6. What percent of the squares below are shaded?

_____

7. What percent of the squares are shaded?

_____

PART 3 • Extended Response

 Record your answer in the space provided.

8. Explain a method you can use to write $\frac{11}{20}$ as a percent.

_____

_____

Lesson 27.2

PART 1 • Multiple Choice

Choose the best answer.

1. How is 70% written as a fraction in simplest form?

   A. $\frac{70}{1}$ C. $\frac{7}{100}$

   B. $\frac{7}{10}$ D. $\frac{7}{100}$

2. How is 63% written as a decimal?

   F. 63.0   H. 0.63

   G. 6.3   I. 0.0063

3. What value of n completes the number sentence correctly?

   $36\% = \frac{n}{25}$

   A. 9   C. 36

   B. 11   D. 61

4. It rained on 6 of the 30 days in November. On what percent of the days did it rain?

   F. 60%   H. 6%

   G. 20%   I. 2%

5. How is 55% written as a decimal and as a fraction in simplest form?

   A. 55 and $\frac{55}{100}$

   B. 55 and $\frac{11}{20}$

   C. 0.55 and $\frac{55}{100}$

   D. 0.55 and $\frac{11}{20}$

PART 2 • Short Response

Record your answers in the space provided.

6. The tax rate in Alpine Lake is 6%. How is this written as a decimal?

   _____

7. What value of x completes the number sentence correctly?

   $85\% = \frac{17}{x}$

   _____

PART 3 • Extended Response

Record your answer in the space provided.

8. One team won 15 of the 25 games it played this season. A second team won 14 of the 20 games it played. Which team won a greater percent of its games? Show your reasoning.

   _____

   _____

Lesson 27.3

PART 1 • Multiple Choice

Choose the best answer.

1. What is 358% written as a decimal?

   A. 35.8

   B. 3.58

   C. 0.358

   D. 0.0358

2. What is 135% written as a fraction or mixed number in simplest form?

   F. $2\frac{3}{5}$

   G. $2\frac{7}{20}$

   H. $1\frac{7}{20}$

   I. $\frac{20}{27}$

3. What is 4.8 written as a percent?

   A. 480%

   B. 48%

   C. 4.8%

   D. 0.48%

4. What is $3\frac{2}{5}$ written as a percent?

   F. 32.5%

   G. 34%

   H. 325%

   I. 340%

5. What is 62.5% written as a fraction?

   A. $\frac{5}{8}$

   B. $\frac{7}{8}$

   C. $1\frac{3}{5}$

   D. $6\frac{1}{4}$

PART 2 • Short Response

Record your answers in the space provided.

6. With the bonus, Alma got 125% on her last test. What is her percent written as a decimal?

   _____

7. To make decisions by voting, many governing bodies use a simple majority which is a number over half. What is the first whole percent over half? Explain.

   _____

   _____

PART 3 • Extended Response

Record your answer in the space provided.

8. The population of a town is 295% of what it was last year. Did the population of the town more than double in size? Explain.

   _____

   _____

Lesson 27.4

PART 1 • Multiple Choice

Choose the best answer.
The table below shows the members of sports teams at Lincoln Elementary School. Use the table for problems 1–6.

1. What percent of the students in the sports program are girls?

Students in Sports Program

	Grade 3	Grade 4	Grade 5	Grade 6
Boys	8	14	13	14
Girls	12	11	12	16
Total	20	25	25	30

   A. 51%   C. 20%

   B. 50%   D. 12%

2. What percent of the Grade 5 students in the sports program are boys?

   F. 88%   H. 48%

   G. 52%   I. 13%

3. In which grade is the percent of boys in the sports program greatest?

   A. Grade 3   C. Grade 5

   B. Grade 4   D. Grade 6

4. In which grade is the percent of girls in the sports program greatest?

   F. Grade 3   H. Grade 5

   G. Grade 4   I. Grade 6

5. What fraction of the students in the sports program are in Grade 3?

   A. $\frac{1}{3}$ C. $\frac{1}{5}$

   B. $\frac{1}{4}$ D. $\frac{1}{6}$

PART 2 • Short Response

Record your answers in the space provided.

6. Tina is in the grade that has the fewest girls in the sports program. What percent of her grade's students are girls?

_____

7. Ross took 10 sample tests to prepare for an exam. The table below shows his work on those tests.

Date	May 1	May 2	May 3	May 4	May 5	May 6	May 7	May 8
Number of Questions	100	50	50	25	25	25	50	50
Number Correct	82	40	42	22	21	23	44	45

On which day did Ross get the highest percent of questions correct?

_____

PART 3 • Extended Response

Record your answer in the space provided.

8. Describe a method you can use to order the numbers 0.45, $\frac{2}{5}$ and 48% from greatest to least.

_____

_____

Lesson 28.1

PART 1 • Multiple Choice

Choose the best answer.

1. What is 40% of 60?

   A. 2.4   C. 24

   B. 4   D. 40

2. What is 110% of $0.70?

   F. $0.07

   G. $0.77

   H. $0.80

   I. $0.81

3. Of the 28 players on a baseball team, 25% are pitchers. How many players are pitchers?

   A. 7   C. 9

   B. 8   D. 25

4. What value of n completes the sentence below correctly?

   35% of n is 63

   F. 18

   G. 22.05

   H. 98

   I. 180

5. What is 250% of 34?

   A. 51

   B. 85

   C. 284

   D. 850

PART 2 • Short Response

Record your answers in the space provided.

6. Of the 1,300 students in a school, 45% live within 1 mile of the school. How many students live within 1 mile of the school?

   _____

7. What value of n completes the sentence below correctly?

   n is 15% of 240

   _____

PART 3 • Extended Response

 Record your answer in the space provided.

8. A sewing machine is regularly priced at $480. The sewing machine is on sale for 20% off its regular price. In addition, there is a 5% sales tax. Show how to calculate the price you would have to pay for this sewing machine.

   _____

   _____

   _____

Lesson 28.2

PART 1 • Multiple Choice

Choose the best answer.

1. What percent is 14 of 70?

    A. 2%   C. 14%

    B. 7%   D. 20%

2. What percent of 160 is 48?

    F. 300%   H. 30%

    G. 33%   I. 3.3%

3. Thirty-five is what percent of 20?

    A. 175%   C. 57%

    B. 125%   D. 20%

4. Which statement is true?

    F. 40% of 40   50% of 20

    G. 30% of 40   20% of 80

    H. 60% of 20   30% of 90

    I. 10% of 50   30% of 60

5. What is the missing value needed below to make the sentence true?

    ☐ is 70% of 90

    A. 0.16   C. 16

    B. 0.63   D. 63

PART 2 • Short Response

Record your answers in the space provided.

6. A librarian saved 150 out of 200 books that were damaged by water. What percent of 200 is 150?

    _____

7. What is the missing value needed below to make the sentence true? Explain.

    80 is ☐% of 50

    _____

    _____

PART 3 • Extended Response

 Record your answer in the space provided.

8. Thad read 35 out of 50 pages and Charles read 80 out of 200 pages. Who read a larger percentage of their book? Explain.

    _____

    _____

    _____

Lesson 28.3

PART 1 • Multiple Choice

Choose the best answer.

1. Of the 15 children in a neighborhood, 10 like to ride bicycles and 8 like to skate. Three children like to do both. How many children like to skate but do not like to ride bicycles?

   A. 2   C. 8

   B. 5   D. 11

2. Of 30 runners surveyed, 22 go running in the morning and 15 go running in the evening. Seven of those surveyed go running in both the morning and the evening. How many of those surveyed go running only in the morning?

   F. 15   H. 29

   G. 22   I. 30

There are 25 students in Jean's class. So far, 14 have signed up for volleyball and 11 have signed up for basketball. Four students have signed up for both sports. Use this information for problems 3–4.

3. How many students have signed up for volleyball but not for basketball?

   A. 10   C. 12

   B. 11   D. 13

4. How many students have signed up for basketball but not for volleyball?

   F. 1   H. 7

   G. 10   I. 5

PART 2 • Short Response

Record your answers in the space provided.

5. The school newspaper staff has writers and editors. There are 12 writers and 6 editors. Two students serve as both writers and editors. How many students are on the staff?

   _____

6. A state has 28 parks with cabins and 56 parks with campsites. Of these parks, 21 have both cabins and campsites. How many parks have cabins but no campsites?

   _____

PART 3 • Extended Response

 Record your answer in the space provided.

7. A survey asked 100 people whether they get their news from the newspaper or from television. Thirty named the newspaper, 82 named television, and 12 named both. Draw a Venn diagram to model this situation. Then use your diagram to identify the number that get their news from each source alone.

   _____

   _____

Lesson 28.4

PART 1 • Multiple Choice

Choose the best answer.
Two hundred students were asked to name their favorite sport. The circle graph shows the results of this survey. Use the circle graph for problems 1–5.

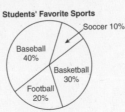

Students' Favorite Sports

Soccer 10%
Baseball 40%
Basketball 30%
Football 20%

1. Which sport was selected by the fewest students?

   A. Baseball   C. Football

   B. Basketball   D. Soccer

2. Which sport was selected by the most students?

   F. Baseball   H. Football

   G. Basketball   I. Soccer

3. How many students named football as their favorite sport?

   A. 20   C. 50

   B. 40   D. 72

4. How many students named basketball as their favorite sport?

   F. 30   H. 60

   G. 50   I. 108

5. How many degrees are there in the part of the circle graph that represents soccer?

   A. 36   C. 20

   B. 25   D. 10

PART 2 • Short Response

Record your answers in the space provided. Students were asked to name their favorite type of movie. The data is shown below. Use the data for problems 6 and 7.

Students' Favorite Types of Movies	Percent of Responses
Action	40%
Comedy	50%
Drama	10%

6. If 250 students were in the survey, how many students chose drama?

_____

7. Suppose a circle graph were made to show this data. How many degrees would there be in the part of the graph that represents action movies?

_____

PART 3 • Extended Response

THINK
SOLVE
EXPLAIN

Record your answer in the space provided.

8. The data in the table represents the favorite beach activities of 300 people. Make a circle graph to show the data.

Favorite Beach Activities	Percent of Responses
Reading	10%
Sleeping	15%
Swimming	60%
Walking	15%

Practice Test

1. An ant starts at a point with coordinates (3, 2). From this point the ant crawls 4 units to the right and then 2 units up. Which three points describe the path the ant traveled on the graph?

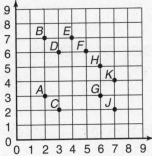

   A. *C* to *J* to *K*

   B. *A* to *G* to *H*

   C. *C* to *D* to *F*

   D. *A* to *B* to *E*

2. Look at the figure below.

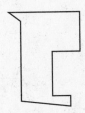

   Which of these figures CANNOT be obtained by reflecting the figure above over a horizontal line or a vertical line?

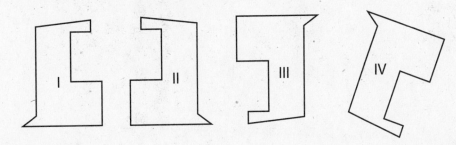

   F. I and III

   G. II and III

   H. II and IV

   J. I and IV

Practice Test

3. Bikes-R-Us charges a fee plus an hourly rate to rent a bicycle. The table below shows the total cost for renting a bicycle for 1 to 5 hours.

Bikes-R-Us					
Number of Hours	1	2	3	4	5
Total Cost	$20	$25	$30	$35	$40

Which equation could be used to represent the total cost (c) for renting a bicycle for h hours?

A. $c = 5 + 15h$

B. $c = 15 + 5h$

C. $c = 20h$

D. $c = 10 + 10h$

4. Evelyn and Jeremy worked together on a project for their history class. Evelyn worked $6\frac{1}{2}$ hours and Jeremy worked $5\frac{3}{4}$ hours. Which expression represents the total number of hours Evelyn and Jeremy worked together on their history project?

F. $6\frac{1}{2} \div 5\frac{3}{4}$

G. $6\frac{1}{2} \times 5\frac{3}{4}$

H. $6\frac{1}{2} - 5\frac{3}{4}$

J. $6\frac{1}{2} + 5\frac{3}{4}$

5. Joe started raising hamsters four years ago. Joe made the graph below showing the number of hamsters he had each year.

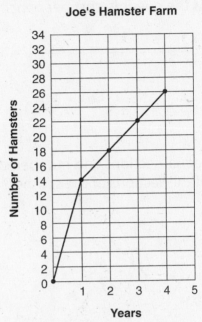

Joe's Hamster Farm

If the pattern continues, how many hamsters will Joe have in the fifth year?

A. 28

B. 30

C. 32

D. 34

6. The graph below shows the results of the last math test in Mrs. Fajardo's class.

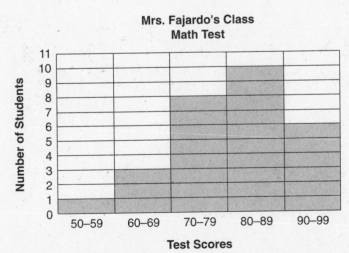

Mrs. Fajardo's Class Math Test

How many students scored 70% or higher on the test? _____

7. Brittany, Victor, and Wesley all bought a box of pencils. Brittany paid $0.11 per pencil and Victor paid $0.15 per pencil. Wesley paid more per pencil than Brittany, but less than Victor. What is the possible cost per pencil that Wesley paid?

_____

8. A spinner is divided into 8 equal sections and numbered as shown below. Which number is on 25% of the sections of the spinner?

A. 1

B. 2

C. 3

D. 4

9. Marge reflected the quadrilateral across the vertical line. She then combined the original quadrilateral with the reflected image. What is the name of the quadrilateral Marge created?

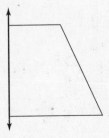

A. rhombus

B. rectangle

C. parallelogram

D. trapezoid

Practice Test

10. Using squares, Paul created the pattern below.

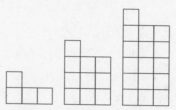

Figure 1   Figure 2   Figure 3   Figure 4

Paul discovered a way of multiplying and subtracting using the figure number to find the total number of squares in any figure. He recorded the pattern in the table below.

Figure Number	Multiply	Subtract	Number of Squares
1	6 × 1 = 6	6 − 2 = 4	4
2	6 × 2 = 12	12 − 2 = 10	10
3	6 × 3 = 18	18 − 2 = 16	16
4			

Draw the next figure in the pattern. Complete the table to find the number of squares in Figure 4.

_____

On the lines below, explain the pattern used to find the answer.

_____

_____

_____

11. Steve made a budget to see how he spends his $20 in allowance each month. The results are shown in the table below.

Budget Category	Percent of Total Allowance
Snacks	20%
Miscellaneous	10%
Entertainment	50%
Savings	20%

Part A

The circle graph on the next page is divided into 10 equal sections. Complete the circle graph to show Steve's budget.

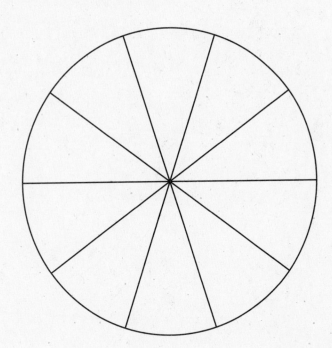

Practice Test

Part B
On the lines below, explain how the circle graph was created.

_____

_____

_____

12. Tina designed a planter in her backyard in the shape shown below.

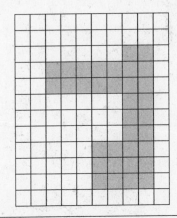

Perimeter = distance around a figure

If the side of a square is 1 foot in length, what is the perimeter of the planter in feet?

_____

13. Mario, Celeste, and Yuri all made art projects using yarn. Mario used $5\frac{1}{2}$ feet, Celeste used 5.6 feet, and Yuri used 5.3 feet. What is the least number of feet of yarn used, written as a decimal?

_____

14. Tara built a patio in the shape shown below using 1-foot square paving bricks.

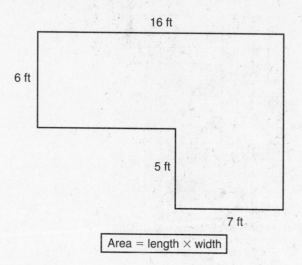

16 ft

6 ft

5 ft

7 ft

Area = length × width

How many paving bricks did Tara use to build the patio?

F. 176

G. 142

H. 131

J. 35

15. Arthur walked the perimeter of his house and counted 84 steps. If each of his steps is 2.5 feet, what is the perimeter of his house, in yards?

1 yard = 3 feet

A. 630

B. 100.8

C. 89.5

D. 70